Research Notes in Neural Computing

Yi-Tong Zhou   Rama Chellappa

Artificial Neural Networks for Computer Vision

With 61 Illustrations

Springer-Verlag
New York Berlin Heidelberg London Paris
Tokyo Hong Kong Barcelona Budapest

Yi-Tong Zhou
HNC, Inc.
5501 Oberlin Drive
San Diego, CA 92121, USA

Rama Chellappa
Department of Electrical Engineering
Center for Automation Research
   and Institute for Advanced Computer
   Studies
University of Maryland
College Park, MD 20742, USA

Managing Editor
Bart Kosko
Department of Electrical Engineering
Signal and Image Processing Institute
University of Southern California
Los Angeles, CA 90089-2564, USA

Library of Congress Cataloging-in-Publication Data
Zhou, Yi-Tong.
    Artificial neural networks for computer vision / Yi-Tong Zhou,
Rama Chellappa.
        p.   cm. — (Research notes in neural computing : v.)
    Includes bibliographical references and index.
    ISBN-13:978-0-387-97683-9   e-ISBN-13:978-1-4612-2834-9
    DOI:10.1007/978-1-4612-2834-9

    1. Neural networks (Computer science)   2. Computer vision.
I. Chellappa, Rama.   II. Title.   III. Series.
QA76.87.Z48   1992
006.3 — dc20   91-27831

Printed on acid-free paper.

Production managed by Christin R. Ciresi; manufacturing supervised by Robert Paella.
Camera-ready copy provided by the authors.

9 8 7 6 5 4 3 2 1

ISBN-13:978-0-387-97683-9

To my wife, Linghong,
and my daughters, May and Daisy.
Yi-Tong Zhou

To my wife, Vishnu Priya,
and my son, Vivek.
Rama Chellappa

Preface

This monograph is an outgrowth of the authors' recent research on the development of algorithms for several low-level vision problems using artificial neural networks. Specific problems considered are static and motion stereo, computation of optical flow, and deblurring an image. From a mathematical point of view, these inverse problems are ill-posed according to Hadamard. Researchers in computer vision have taken the "regularization" approach to these problems, where one comes up with an appropriate energy or cost function and finds a minimum. Additional constraints such as smoothness, integrability of surfaces, and preservation of discontinuities are added to the cost function explicitly or implicitly. Depending on the nature of the inversion to be performed and the constraints, the cost function could exhibit several minima. Optimization of such nonconvex functions can be quite involved. Although progress has been made in making techniques such as simulated annealing computationally more reasonable, it is our view that one can often find satisfactory solutions using deterministic optimization algorithms.

In this monograph, we present deterministic optimization algorithms using artificial neural networks of the Amari–Hopfield type for several low-level vision problems. For each one of these problems, we have identified appropriate constrained cost functions. For instance, features based on estimated first derivatives and Gabor wavelets are used for defining the cost function used in stereo matching. A cost function using features based on principal curvatures is defined for computation of optical flow. For the motion stereo problem, the cost function used for static stereo is extended to multiple frames, leading to a recursive solution. Subsequently, the results of minimizing these cost functions are presented for several synthetic and real images. Practical issues such as the effects of spatial quantization, detection of occluding pixels, choice of window size used for estimation of derivative, and detection of motion discontinuities are discussed in detail. Thus, the emphasis is on engineering the artificial neural networks for several image-related problems. Some of the algorithms presented in this monograph have already been implemented in VLSI hardware by Professor B. Sheu and his students at the University of Southern California. Optical implementations of image deblurring and stereo matching algorithms have been investigated by Professor B.K. Jenkins and his students. It is our hope that this monograph serves as a practical book for engineers and scientists interested in exploiting the computational power of artificial neural networks for image processing and computer vision problems.

During the last five years, we have been tremendously inspired and influenced by many of our distinguished colleagues at the University of Southern California. In particular, we would like to thank Professor M.A. Arbib, Director of the Center for Neural Engineering, for his unbridled enthusiasm and leadership in neural-related research activities in USC. We would like to acknowledge the profound influence Professor C. von der Malsburg has had on the work reported here and subsequent work the second author has done with Dr. B.S. Manjunath. We would also like to thank Professor B.A. Kosko for his dynamic and thought-provoking interactions and his continued encouragement to complete this work.

We would like to thank Dean L.M. Silverman, Dr. J.M. Mendel, Dr. A.A. Sawchuk, Dr. B.K. Jenkins, Dr. B. Sheu, Dr. A. Weber, Dr. B.S. Manjunath, Ms. Linda Varilla, and Ms. Delsa Tan for their encouragement, helpful discussions, and assistance. The first author would also like to thank Dr. Robert Hecht-Nielsen, Dr. Robert L. North, Mr. Todd Gutshow, Dr. Robert Means, Mr. Richard Crawshaw, Mr. Chris Platt, and Ms. Sherri Mieth of HNC for their advice, support and help. Thanks are also due to the editorial staff at Springer-Verlag for their encouragement and patience during the preparation of this monograph.

Finally, this work would not have been completed without the support of our families.

The research reported in this monograph was supported by the AFOSR Grant 86-0196 and by the Center for Integration of Optical Computing, which was supported by the AFOSR Contract F-49620-87-C-007 and the AFOSR Grant 90-0133.

San Diego, California   Yi-Tong Zhou
Los Angeles, California   Rama Chellappa

Contents

1
Introduction

1.1   Neural Methods

This book is concerned with developing algorithms for some important computer vision problems, especially at a low-level using artificial neural networks. The task of low-level vision is to recover physical properties of visible three-dimensional surfaces from two-dimensional images. One module of low-level vision, for instance, extracts depth information from two eyes, making binocular images, or from one eye over a period of time, making a sequence of monocular images. Low-level processes also provide motion information about objects over a sequence of two-dimensional images, the optical flow, for motion detection and representation. As intelligent interpretation of an image requires knowledge about the objects that appear in the scene, learning, representation, and use of prior knowledge must be coordinated. Certainly the human brain is very good at performing vision tasks, but today's computers are not. This is because of the massive amount of two-dimensional array data that needs to be analyzed and the lack of learning or self-organizing capabilities of most modern day computers. From a mathematical point of view, low-level vision problems are ill-posed according to Hadamard [Had02, Mor84]. An efficient method for solving an ill-posed problem using artificial networks is the Tikhonov regularization technique. The idea of the regularization technique is to narrow the admissible solution region by introducing suitable *a priori* knowledge or stablizing the solution by means of some auxiliary non-negative functional [Mor84]. Cognitive scientists believe that the brain uses cooperative computation to deal with the task of low-level vision [AH87]. Cooperative computation takes into account multiple and often mutually conflicting constraints or many pieces of information simultaneously to narrow the solution region. This is accomplished by employing massively parallel processing units, with information transfer taking place through the interconnections between these units, in the form of excitatory and inhibitory signals. This cooperative computation provides not only massive parallelism, but also a greater degree of robustness because of local connectivity among processing units and adaptability of interconnection strengths (weights). Attempting to provide a human-like performance, many neural network models have been developed based on studies of cooperative computation in biological nervous systems. The neural network models seem well suited for mimicking low-level vision processes and are thus an attractive way of solving the low-level computer vision problems. The neural network models have greatest potential in the low-level vision areas where highly parallel computations are required and the currently best computer systems are inferior to human performance.

In this book, some artificial neural networks are used to solve computer vision problems such as static stereo, lateral motion stereo, longitudinal motion stereo, computation of optical flow, and image restoration. Networks such as these, which containing massive mutually interconnected and self connected binary neurons, are very similar to Amari [Ama71, Ama77] and Hopfield networks [HT85]. Since self-feedback may cause the energy function of the network to increase with a transition, two decision rules, deterministic and stochastic, are used to ensure convergence. The stochastic decision rule guarantees that the network will converge to a global minimum but is computationally intensive. The deterministic decision rule greatly reduces computational time but gives only a local minimum, an approximate solution. Two types of activation functions, the threshold function (step function) and the maximum evolution function, are used in the updating scheme.

Usually, the measurement primitives used for stereo matching are the intensity values, edges and linear features. Conventional methods based on such primitives suffer from amplitude bias, edge sparsity and noise distortion, whereas the human stereo process does not. Knowing that the human visual system is very sensitive to intensity changes, derivatives of the intensity function which are more reliable, dense and robust are used for matching in this study. For estimating the derivatives, a window operator that combines smoothing and differentiation is suggested. However, as the natural stereo images are digitized, the resulting spatial quantization error affects the intensity function and the derivatives. The effects of noise and spatial quantization on the estimation of derivatives are discussed, leading to an appropriate choice of window size. As Gabor functions have been widely used in the vision area, we investigate the possibility of using Gabor features as the measurement primitives for matching. A Gabor function is a sinusoidal function weighted by a Gaussian window. The Gabor features are obtained by convolving the image with a set of Gabor functions. To reduce the computation load, we truncate the Gabor function to a central area with a diameter of several standard deviations of the Gaussian function.

Recovering depth is a central problem in three-dimensional perception. Physiological studies shows that depth can be recovered based on either monocular cues or binocular cues. Static stereo and motion stereo are two basic methods for inferring depth information from multiple images. Static stereo uses one pair of binocular images and motion stereo needs at least two monocular images. Since motion stereo uses more than two image frames, it usually gives more accurate depth measurements than static stereo. To simplify the problem, most static stereo algorithms assume that the baseline is known and that epipolar, photometric and smoothness constraints are satisfied. For the same reason, our algorithm also adopts these assumptions. By using the network to represent a disparity field, stereo matching is carried out by neuron evaluations. Although many researchers have been using neural networks for stereo matching [PD75, MP76, GL86, SD87, SCL87, GG87], their performance on natural images is not well established. Experimental results using both synthetic and natural stereo images show that our algorithm successfully recovers the depth information. The robustness of the algorithm is illustrated by using several pairs of decorrelated images.

Motion stereo uses multiple monocular images to infer depth information. According to the nature of motion, motion stereo can be further divided into three categories: lateral, longitudinal and rotational motion stereo. Existing lateral motion stereo algorithms use either a Kalman filter or recursive least square algorithm [MSK88] to update the disparity values. Due to the unmeasurable estimation error, the estimated disparity values at each recursion are unreliable, yielding a noisy disparity field. Instead of updating the disparity values, our approach recursively updates the bias inputs of the network, the measurement primitives. The disparity field is then computed by using a static matching algorithm based on the neural network. Since the recursive algorithm implements the matching algorithm only once, and the bias input updating scheme can be accomplished in real time, a vision system employing such an algorithm is feasible. A detection algorithm for locating occluding pixels is also included. No surface interpolation and smooth procedures are required in all algorithms. For the purposes of handling batch data, a batch algorithm is also presented. The batch algorithm integrates information from all images by embedding them into the bias inputs of the network. Then the same static matching procedure is used to compute the disparity values. One sequence of natural images is considered in the experimental test. For longitudinal motion stereo, an algorithm based on three frames is described. Existing approaches usually require some information about the location of the focus of expansion (FOE), have some restrictions on the moving direction of the camera and the orientation of the object surface. These requirements and restrictions limit their applications in real scenes. Instead, our approach needs no information about the FOE, allows the camera to move along its optical axis freely, and makes no assumption about the object surface.

Optical flow is the distribution of apparent velocities of moving brightness patterns in an image. Motion can be caused by moving objects and/or a moving camera. The problem considered here is how to calculate the optical flow from two or more image frames. Starting with conventional methods, an algorithm is proposed for computing optical flow from two image frames based on principal curvatures, rotation invariant measurement primitives . Under local rigidity assumption and a smoothness constraint, a multi-layer neural network is then used to compute the optical flow. A difficult problem in computing optical flow is to locate motion discontinuities. A line process is commonly used [Koc87] for detecting discontinuities. However, the detected discontinuities may be shifted due to an occluding region, because the optical flow at the occluding region is undetermined. In order to locate the discontinuities accurately, we first detect the occluding elements based on the initial motion measurements and then embed this information into the bias inputs of the network. Computer simulations on synthetic and natural images are presented. When motion is pure translation with constant velocity, optical flow can be estimated from a sequence of images. A batch solution is discussed for this problem and successfully tested on a sequence of natural images with a stationary background.

Restoration of a high-quality image from a degraded recording is an important problem in early vision processing. An approach for restoration of gray level images degraded by a known shift invariant blur function and additive noise is developed. A sequential neural network is employed to represent a possibly nonstationary image whose gray level function is the simple sum of the neuron state variables. The nonlinear restoration method is carried out iteratively by using a dynamic algorithm to minimize the energy function of the network. Owing to the model's fault–tolerant nature and computational capability, a high quality image is obtained using this approach. A practical algorithm with reduced computational complexity is also presented. Several computer simulation examples involving synthetic and real images are given to illustrate the usefulness of our method. To reduce the ringing effect, the choice of the boundary values is discussed. Comparisons with other restoration methods such as the SVD pseudoinverse filter, minimum mean square error (MMSE) filter and the modified MMSE filter using the Gaussian Markov random field model are given. An optical implementation of this approach is also described.

1.2   Plan of the Book

This book is made up of eight chapters. Each chapter (except Chapters 1 and 7) begins with an introduction that provides an overview of the topic and presents the motivation for the approach, and ends with a discussion that gives our thoughts about promising future research directions.

Chapter 2 reviews the relevant artificial neural networks and gives a modified neural network, which is used in this research. Two decision rules, deterministic and stochastical, are presented.

In Chapter 3, static stereo research is reviewed and a neural network-based algorithm using epipolar, photometric and smoothness constraints is discussed. A window operator for estimation of the first derivatives of intensity functions is derived. An analysis of the effects of noise and spatial quantization on the estimation of the derivatives is given. Results from synthetical and natural images are presented.

Chapter 4 addresses the motion stereo problem where a monocular motion image sequence is used instead of binocular images. The motion images are assumed to be taken by a laterally moving camera. After briefly reviewing the existing literature, two approaches, known as the batch and recursive approaches, are proposed for lateral motion stereo. Detection of occluding pixels based on the matching error is also discussed. One sequence of real motion images is used to illustrate the performances of these approaches.

Chapter 5 continues to discuss the motion stereo problem, although only the longitudinal motion case is considered. Existing algorithms are first reviewed. A new longitudinal motion stereo algorithm is then presented. Unlike the static and lateral motion stereo algorithms, Gabor features are used in the longitudinal motion stereo as the measurement primitives for matching. One section is devoted to designing the Gabor feature extractor. The chapter concludes with a set of test results on the real image data.

Chapter 6 deals with optical flow. Existing methods for computing optical flow are discussed. A method for finding the principal curvatures based on the second order derivatives estimated with subpixel accuracy is suggested. A neural network algorithm for computing optical flow from two successive image frames is developed and extened to the case of multiple sequential images. A method for detecting motion discontinuities is derived, and experimental results using synthetic and natural images are presented.

Chapter 7 presents an approach for image restoration. By using a simple sum number representation, a dynamic algorithm using the neural network is developed. A practical algorithm with reduced computational complexity is also presented. Discussions on the choice of boundary values and comparisons to other restoration methods are given. Several computer simulation examples are presented to show the usefulness of this method. An optical implementation of this approach is presented.

Chapter 8 concludes the book with an overview of current neural network research in computer vision, and highlights some key unresolved issues and directions for future research.

2

Computational Neural Networks

2.1   Introduction

Research on neural network modeling has a long history. Neurobiologists have discovered individual nerve cells existing in the brain and learned how neurons carry information, transmit information, and respond to various stimuli. Based on the understanding of the nervous system, many neural networks have been proposed by researchers. Over the past fifty years, thousands of papers have been published in this area. As early as 1943, Mc-Culloch and Pitts [MP43] developed a neural network by treating neurons as Boolean devices and showed that such a network could compute. This network used a step function as the activation function which has been adopted by many neural networks such as the Amari recurrent network [Ama71, Ama77], the discrete Hopfield network [Hop82], and the discrete bidirectional associative memory [Kos88]. Recently, learning has become the main focus in this area. In 1949, Hebb [Heb49] proposed a learning rule that is a simulated network; first tested in the Edmonds and Minsky's learning machine, it is still used today in many learning paradigms. In the 1950s, Rosenblatt [Ros59, Ros62] invented a class of simple neuron learning networks called perceptrons in order to realize a dynamic, interactive and self-organizing system. Minsky and Papert [MP69] studied Rosenblatt's learning networks and found that a two-layer network would only work for the linear separatable problems. Meanwhile, Selfridge [Sel55] developed a dynamic, interactive network for computational tasks in perception. Widrow [Wid59] created a two-layer adaptive linear element called ADALINE similar to the perceptron. Widrow's adaptive network used a delta learning rule for pattern recognition [WH60]. Since its development, ADALINE has been successfully used for adaptive signal processing [WS85]. Anderson [And77] provided brain state in a box model which contains a feedback loop for learning. Recently, Amari [Ama72, AA77], Arbib [Arb64, AA77], von der Malsburg [vdM73], Fukushima [Fuk75], and Grossberg [Gro76] have developed several competitive learning models. Also Kohonen [Koh77], Feldman and Ballard [FB82], Rumelhart and McClelland [RM82], Hopfield[Hop82, Hop84, HT85], Sejnowski [SR86], and others have made important contributions to neural network modeling.

In 1982, Hopfield [Hop82] developed a neural network and presented a method to analyze the stable point of the neural network. His method is based on the Lyapunov energy function. Later, Hopfield extended his original model from discrete to continuous time [Hop84]. In 1985, Hopfield and Tank [HT85] showed that a certain class of optimization problems can be programmed and solved on their neural network. Since then the computational power of the neural network has become more and more apparent. The neural networks used in this research are based to a great extent on the Amari recurrent network and the Hopfield discrete network.

2.2 Amari and Hopfield Networks

Amari has studied neural networks since the late 1960s. In the early 1970s, Amari [Ama71, Ama77] proposed two self-organizing random networks: a non-recurrent network for association and a recurrent network for concept formation. The recurrent network, shown in Figure 2.1(a), is a sequential network containing n bistable elements (neuron pools) $\{v_1(t), v_2(t), ..., v_n(t)\}$. Each element, shown in Figure 2.1(b), consists of mutually connected neurons. The outputs of the network are connected to its inputs. The bistable element can be in one of two states: $v_i(t) = 1$ (firing state) if many neurons in the pool are active and $v_i(t) = 0$ (resting state) if many neurons in the pool are off. Each element $v_i(t)$ receives weighted input signals from all the elements at time t. After summing the weighted inputs and comparing with a threshold h_i, the state of the element $v_i(t+1)$ at time $t+1$ is determined by

$$v_i(t+1) = g(\sum_{j=1}^{n} T_{i,j}\, v_j(t) - h_i) \qquad (2.1)$$

where $T_{i,j}$ is the weight (synaptic interconnection strength) from element j to element i and $g(x)$ is the activation function defined by

$$g(x) = \begin{cases} 1 & if \ x \geq 0 \\ 0 & if \ x < 0, \end{cases} \qquad (2.2)$$

which is a step function. The stable states are reached if

$$v_i(t+1) = v_i(t) \quad for \quad i = 1, 2, ..., n.$$

The pattern specified by the stable states of the elements is the network output. Since the weight $T_{i,i}$ is nonzero, all the elements of the network have feedback. The recurrent network was actually derived from the McCulloch-Pitts formal neuron, the simplest form of neural network [MP43].

The Hopfield discrete network, shown in Figure 2.2, uses two state threshold neurons $\{v_1(t), v_2(t), ..., v_n(t)\}$. Each neuron receives external input I_i and weighted inputs from other neurons at random times. The total input

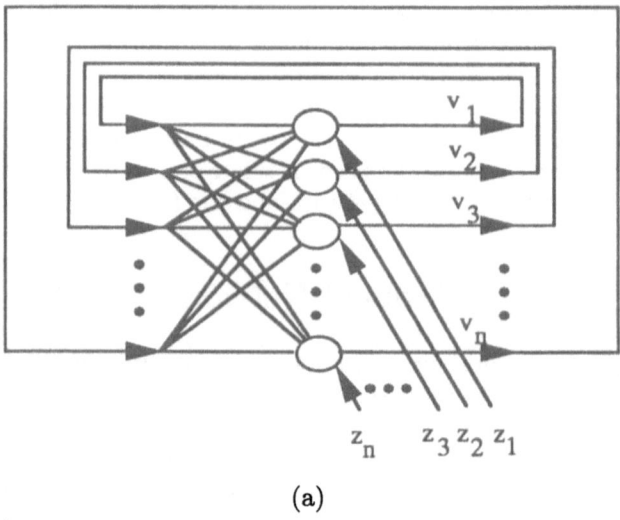

(a)

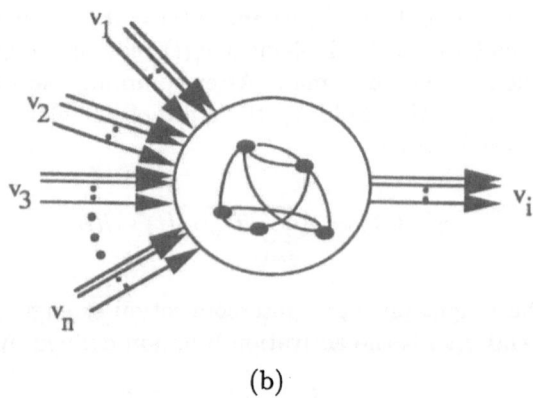

(b)

FIGURE 2.1. Amari recurrent network. (a) Recurrent network. (b) Bistable element (neural pool).

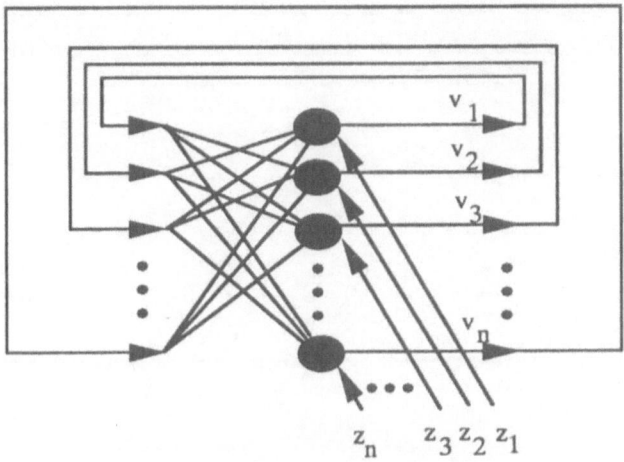

FIGURE 2.2. Hopfield discrete network.

of neuron i is

$$\sum_{j \neq i} T_{i,j}\, v_i(t) + I_i.$$

The state of the neuron i is asynchronously updated according to a threshold rule with threshold h_i

$$v_i = g(\sum_{j=1}^{n} T_{i,j}\, v_j(t) + I_i - h_i) \qquad (2.3)$$

where $g(x)$ is defined in (2.2). The asynchronous property is introduced to represent a combination of propagation delay, jitter and noise in real systems. To ensure that the network converges to stable states, two conditions, symmetric interconnections ($T_{i,j} = T_{j,i}$) and no self-feedback ($T_{i,i} = 0$), have to be satisfied [Hop84]. If we consider the bistable elements of the Amari network as the binary neurons, one can see that the Hopfield network is similar to the Amari network. The Hopfield network is originally designed for association. However, Hopfield and Tank [HT85] used an analog version of this network to solve a difficult but well defined optimization problem (the Traveling Salesman problem). Recently, they have also used the neural network for speech recognition [TH87].

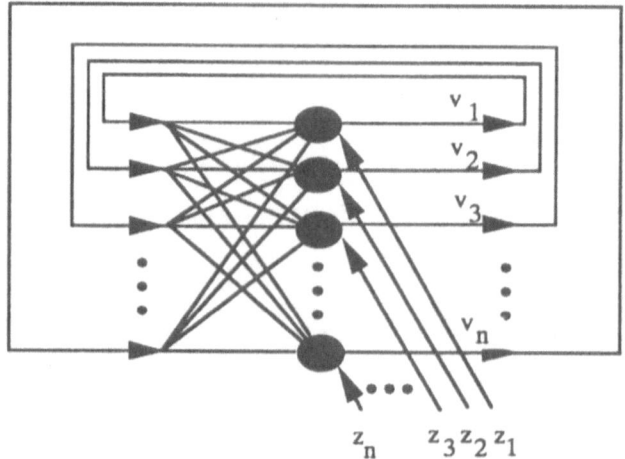

FIGURE 2.3. A discrete neural network.

2.3   A Discrete Neural Network for Vision

To deal with vision problems, we use several multi-dimensional neural networks. For illustration purposes, we present only a one-dimensional neural network here. More details about the multi-dimensional networks will be given in subsequent chapters. It should be noted that since images are two-dimensional data, even if we use one neuron to represent each image pixel, then a huge number of neurons are needed to represent the whole image. For instance, for a 256×256 image, a total of $65,536$ neurons are needed. If each pixel needs multiple neurons, say m neurons, then a total of $65,536 \times m$ neurons are required. Figure 2.3 shows our 1-D neural network.

2.3.1   A DISCRETE NETWORK

The network consists of n mutually interconnected binary neurons $\{v_1, v_2, ..., v_n\}$. Each neuron takes the value 0 for resting and 1 for firing. Let $T_{i,j}$ denote the strength (possibly negative) of the interconnection between neuron i and neuron j. The interconnections are assumed to be symmetric

$$T_{i,j} = T_{j,i} \quad \text{for} \quad 1 \leq i,j \leq n$$

and the self-connection is not necessarily zero. the self-connection could be nonzero

$$T_{i,i} \neq 0.$$

The nonzero self-connection means there is a self-feedback for each neuron. In this network, each neuron (i, k) synchronously, or randomly and asynchronously receives inputs $\sum T_{i,j} v_j$ from all neurons including itself and a bias input I_i:

$$u_i = \sum_j^n T_{i,j} v_j + I_i. \tag{2.4}$$

Each neuron u_i is fed back to corresponding neurons after either thresholding or maximum evolution

$$v_i = g(u_i) \tag{2.5}$$

where $g(x_i)$ is an activation function whose form is taken either as (2.2) for thresholding or

$$g(x_i) = \begin{cases} 1 & if\ x_i = max(x_k; \forall k \in \Omega_l) \\ 0 & otherwise \end{cases} \tag{2.6}$$

for maximum evolution (winner-take-all), where Ω_l's are disjoint subsets of index set $\Omega = \{1, 2, ..., n\}$ and $\bigcup \Omega_l = \Omega$, and $i \in \Omega_l$. The synchronous updating scheme uses information about the old states of all the neurons. By contrast, the asynchronous updating scheme uses the latest information about the states of the other neurons to update the state of the present neuron, which means that any state change in a neuron will immediately affect the state of all the neurons.

2.3.2   DECISION RULES

As mentioned above, this network has a self-feedback, $T_{i,i} \neq 0$. The reason for having the self-feedback will be given in subsequent chapters. As a result of having the self-feedback, this network does not always converge to stable states. This can be explained as follows. Let E denote the energy function of the network. According to [Hop84], by setting thresholds $\{h_i\}$ to zero the energy function of the network can be found as

$$E = -\frac{1}{2} \sum_{i=1}^n \sum_{j=1}^n T_{i,j}\, v_i\, v_j - \sum_{i=1}^n I_i\, v_i. \tag{2.7}$$

Let the state change of neuron i be

$$\Delta v_i = v_i^{new} - v_i^{old}$$

and the energy change of the network be

$$\Delta E = E^{new} - E^{old}.$$

Case 1: Step function.

For simplicity of analysis, we assume that at each step, only one neuron changes its state either from 1 to 0 or from 0 to 1. When a step function is used as the activation function, the energy change ΔE due to a state change Δv_i of neuron i is given by

$$\Delta E = -(\sum_{j=1}^{n} T_{i,j}\, v_j + I_i\,)\, \Delta v_i - \frac{1}{2}\, T_{i,i}\, (\Delta v_i)^2. \qquad (2.8)$$

By (2.4), ΔE can be written as

$$\Delta E = -u_i\, \Delta v_i - \frac{1}{2}\, T_{i,i}\, (\Delta v_i)^2. \qquad (2.9)$$

When u_i is greater than zero, v_i changes its state from 0 to 1 and hence $\Delta v_i = 1$ which leads to

$$\Delta E = -u_i - \frac{1}{2}\, T_{i,i}. \qquad (2.10)$$

If $T_{i,i} < -2u_i$, then $\Delta E > 0$. Similarly, when u_i is less than 0, v_i changes its state from 1 to 0. The state change Δv_i is then -1 and

$$\Delta E = u_i - \frac{1}{2}\, T_{i,i}. \qquad (2.11)$$

If $T_{i,i} < 2u_i$, then $\Delta E > 0$. Hence, whenever

$$T_{i,i} < -2|u_i|,$$

we have $\Delta E > 0$ which means that the energy changes are not always negative and the energy function does not decrease monotonically with a transition. E is not a Lyapunov function and the network is unstable. Consequently, the convergence of the network is not guaranteed [LaS86].

Case 2: Maximum evolution function.

When a maximum evolution function is used, a batch of m neurons $\{v_k; k \in \Omega_l\}$ is simultaneously updated at each step. Since the maximum evolution function only allows one neuron to be active and the others to be inactive, at most two neurons can change their state at each step, the active neuron becomes inactive and one of the inactive neurons becomes active. Suppose neurons i and i' change their states. The energy change ΔE due to the state changes of neurons i and i' is then given by

$$\begin{aligned}
\Delta E \;=\; & -(\sum_{j=1}^{n} T_{i,j}\, v_j + I_i\,)\, \Delta v_i - \frac{1}{2}\, T_{i,i}\, (\Delta v_i)^2 \\
& -(\sum_{j=1}^{n} T_{i',j}\, v_j + I_{i'}\,)\, \Delta v_{i'} - \frac{1}{2}\, T_{i',i'}\, (\Delta v_{i'})^2 \\
& -T_{i,i'}(\Delta v_i\, v_{i'}^{new} + \Delta v_{i'}\, v_i^{new}). \qquad (2.12)
\end{aligned}$$

Similarly, by (2.4), ΔE can be written as

$$\Delta E = -u_i \, \Delta v_i - \frac{1}{2} T_{i,i} \, (\Delta v_i)^2 - u_{i'} \, \Delta v_{i'} - \frac{1}{2} T_{i',i'} \, (\Delta v_{i'})^2$$
$$-T_{i,i'}(\Delta v_i \, v_{i'}^{new} + \Delta v_{i'} \, v_i^{new}). \tag{2.13}$$

By properly setting v_i and $v_{i'}$, it is easy to show that the energy changes are not always negative. More details about how the energy change can be made positive are given in Chapter 3.

To ensure convergence of the network to a minimum, one can design some decision rules for updating the states of neurons. Depending on whether convergence to a local minimum or a global minimum is desired, a deterministic or stochastic decision rule can be used, respectively. In some cases, for example when the energy function is convex, the deterministic decision rule will ensure that the network will converge to a global minimum.

Deterministic Decision Rule:
The deterministic rule is to take a new state v_i^{new} of neuron i if the energy change ΔE due to state change Δv_i is less than zero. If ΔE due to the state change is > 0, no state change is affected.

Stochastic Decision Rule:
A stochastic rule is similar to the one used in simulated annealing techniques [MRR+53, KGV83, GG84]. Define a Boltzmann distribution by

$$\frac{p_{new}}{p_{old}} = e^{\frac{-\Delta E}{T}}$$

where p_{new} and p_{old} are the probabilities of the new and old global state respectively, ΔE is the energy change and T is the parameter which acts like temperature. A new state v_i^{new} is taken if

$$\frac{p_{new}}{p_{old}} > 1, \quad or \; if \;\; \frac{p_{new}}{p_{old}} \le 1 \;\; but \;\; \frac{p_{new}}{p_{old}} > \xi$$

where ξ is a random number uniformly distributed in the interval $[0,1]$.

2.4 Discussion

In this chapter, we have presented a discrete artificial neural network. Due to self-feedback , two decision rules have been suggested to ensure convergence of the network. Comparing this network with the Amari network and Hopfield network, some differences listed in Table 2.1 can be noted.

The difference among them is that the network used in this book has self-feedback and hence requires more effort to ensure convergence. Self-feedback arises naturally in the problems considered in this book. For the

TABLE 2.1. Comparisons to the Amari network and Hopfield network.

Network	Neuron	Self-feedback	Decision rule	Activation function
Amari	Binary, pool	Yes	Deterministic	Step
Hopfield	Binary	No	Deterministic	Step
Ours	Binary	Yes	Deterministic, Stochastic	Step, Maximum evolution

Traveling Salesman problem [HT85], based on simulations we have found that our network needs fewer iterations than the Hopfield network. Apparently, using the deterministic decision rule in a network with self-feedback results in fewer iterations than are required by a network without self-feedback.

3

Static Stereo

3.1 Introduction

Recovering depth is a central problem in three-dimensional perception. Static stereo is a primary means for recovering depth from two images taken from different viewpoints. As early as 1838, Sir Charles Wheatstone [Whe38] invented a stereoscope which uses the slight differences of two static pictures to generate a vivid sense of depth. This invention is based on the phenomenon of stereopsis, which arises from the fact that human eyes view the visual world from slightly different angles. As shown in Figure 3.1, when an observer looks at a point A in space, the images of A projected on the centers of the left and right foveas are L_A and R_A, respectively. L_A and R_A are called corresponding points. Now suppose there is another point B in space which is farther from the observer than point A. Point B produces two images L_B and R_B on the retinas some distance from the centers of the foveas. The distances from L_B to the center of the left fovea and from R_B to the center of the right fovea are different. The disparity in the distance varies with the depth of the point in space. The three-dimensional information can be decoded from the binocular disparities.

A large number of static stereo methods exists. Most methods, however, fall into one of two categories: region-based and feature-based, according to the nature of the measurement primitives. The region-based methods use the intensity values as the measurement primitives. A correlation technique or some simple modification is applied to certain local region around the pixel to evaluate the quality of matching. The region-based methods usually suffer from problems due to lack of local structures in homogeneous regions, amplitude bias between the images, and noise distortion. Recently, Barnard [Bar86] has applied a stochastic optimization approach for the stereo matching problem to overcome the difficulties due to homogeneous regions and noise distortion. Although this approach is different from the conventional region-based methods, it still uses intensity values as the primitives with the aid of a smoothness constraint. Barnard's approach has several advantages: simple, suitable for parallel implementation, and a dense disparity map output. However, too many iterations, a common problem with the simulated annealing method, makes it unattractive. It also suffers from the problems of amplitude bias between the two images and oversmoothing.

The feature-based methods use intensity edges, linear features (for example, see Grimson [Gri81] and Medioni [MN85]), or intensity peaks which cor-

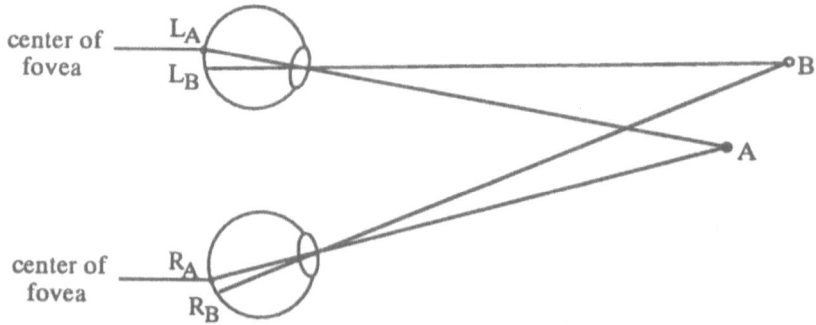

FIGURE 3.1. Principle of stereopsis.

respond to discontinuities in the first order derivatives of intensity [MF81].
The intensity edges are obtained using edge detectors such as the Marr–
Hildreth edge detector [MH80] or the Nevatia–Babu line finder [NB80].
Since amplitude bias and small amounts of noise do not affect edge detection, feature-based methods can handle natural images more efficiently.
Owing to fewer measurement primitives to deal with, feature-based methods are usually faster than the region-based methods. However, a surface
interpolation step has to be included. In order to obtain a smooth surface, several types of smoothness constraint techniques have been introduced [Gri81]. The common problem in feature-based methods is that if
features are sparse, then the surface interpolation step is difficult. Among
the feature-based methods, the Marr-Poggio-Grimson algorithm gives impressive results. But it is difficult to ensure continuity of disparity over an
area of the image. To overcome this problem, Grimson [Gri85] proposed a
new method including the figural continuity constraint [MF81] and other
modifications. The figural continuity constraint is superior to the region
continuity constraint. However, an occluding boundary or a sloping surface may cause problems. Another interesting approach is the integrated
approach, which combines matching, contour detection and surface interpolation steps [HA87]. The integrated approach uses only a piecewise smoothness assumption. A number of stereo images were tried in [HA87] to illustrate the performance of this approach. Some problems of this approach
reported by the authors are misplacement and deletion of contours, and
disparity errors due to inaccuracies in edge detection.

Julesz'sindexJulesz example of random dot stereograms shows that stereo
matching occurs very early in the visual process and is relatively independent of other forms of visual processing [Jul71]. The early visual process implies that more dense measurement primitives are used in stereo
matching. It seems that the region-based methods are closer to the hu-

man stereo process than the edge-based methods, because the intensity values are dense measurement primitives. However, region-based methods suffer from the problems of amplitude bias and noise distortion, whereas the human stereo process does not. The question, then, is what sort of measurement primitives the human stereo process does use. Arguing that the amplitude bias can be eliminated by the differential operation, the intensity derivatives are dense, and that the human visual system is sensitive to the intensity changes, the first-order intensity derivatives (the simplest derivatives) may be considered to be appropriate measurement primitives for the stereo matching problem. Noise distortion, which the first order derivatives are very sensitive to, can be reduced by some smoothing techniques, such as polynomial fitting. The first order intensity derivatives can be obtained by directly taking the derivative of the resulting continuous intensity function. Actually, the choice of window size is closely related to the theory of the human visual system. There exist at least four independent channels containing different sized spatial filters in the early visual system [WG77, WB79]. A combination of smoothing and differentiation results in a window operator which functions very much like the human eye in detecting intensity changes. To obtain some insights into the resulting window operator, a theoretical analysis of the variances of the estimated derivatives is given. Since the natural stereo images are usually digitized for implementation on a digital computer, we consider the effect of spatial quantization on the estimation of the derivatives from natural images.

Recently, many researchers have been using neural networks based on either intensity values or edges [GL86, SD87, SCL87, GG87] for stereo matching. Early work on extracting depth information from the random dot stereogram using a neural network may be found in [PD75, MP76]. In [MP76], a cooperative algorithm is employed to compute correspondence between the two descriptions, subject to uniqueness and continuity. Unlike standard correlation techniques, this algorithm is not restricted to minimum or maximum correlation areas to which the analysis is subsequently restricted. Although the algorithm is based on primitive descriptions such as edges for matching, no preprocessing procedure was involved in their experiments because they considered each white dot in the binary random dot stereogram as a primitive. Extension of this algorithm to natural images was reported in [Pog84]. The natural images are first converted into binary maps by taking the sign of their convolution with the Laplacian of a Gaussian. Then the resulting binary maps serve as inputs for the cooperative algorithm. Grimson [Gri81] further extended this algorithm using zero crossings.

In this chapter, we present a neural network algorithm for deriving depth information from two views. First, two cameras are set at two different places for taking pictures. Then, a filter is used to extract the measurement primitives from images. Finally, a neural network is employed to find the conjugate points and compute the disparity values based on measurement primitives under the epipolar, photometric and smoothness constraints. The usefulness of the algorithm is illustrated by using the random dot stereograms and natural image pairs.

3.2   Depth from Two Views

Two cameras used for stereo matching are located as shown in Figure 3.2. These cameras are rigidly attached to each other so that their optical axes are parallel and separated by a distance d. The focal length of the lens is denoted by f which takes a negative value in the space coordinate system $OXYZ$. The origin of the right-handed coordinate space system is located midway between the camera lens centers. The positive Z-axis is directed along the camera optical axes. The baseline connecting the lens centers is assumed to be perpendicular to the optical axes and oriented along the Y-axis. The coordinate systems of the left and right image planes are given by $o_L x_L y_L$ and $o_R x_R y_R$, respectively. Now suppose a point P in space (X_P, Y_P, Z_P) projects into the left and right image planes at (x_{PL}, y_{PL}, z_{PL}) and (x_{PR}, y_{PR}, z_{PR}), respectively. By similar triangles, we have

$$\frac{y_{PL}}{f} = \frac{Y_P + \frac{d}{2}}{Z_P} \tag{3.1}$$

$$\frac{y_{PR}}{f} = \frac{Y_P - \frac{d}{2}}{Z_P}. \tag{3.2}$$

The image point positions are then determined by

$$(x_{PL}, y_{PL}) = (\frac{f\,X_P}{Z_P}, \frac{f\,(Y_P + \frac{d}{2})}{Z_P}) \tag{3.3}$$

and

$$(x_{PR}, y_{PR}) = (\frac{f\,X_P}{Z_P}, \frac{f\,(Y_P - \frac{d}{2})}{Z_P}). \tag{3.4}$$

Let $D_{X,Y}$ denote the disparity value in the y direction. By subtracting y_{PL} from y_{PR}, we have

$$D_{X,Y} \triangleq y_{PR} - y_{PL}. \tag{3.5}$$

As the distance is inversely proportional to the disparity

$$Z_P = -\frac{f\,d}{D_{X,Y}},$$

it is straightforward to calculate the distance from the disparity when the baseline and the focal length are known. Stereo matching usually computes the disparity values only. The central problems in stereo matching are (1) extracting and matching corresponding feature points or lines, and (2) obtaining a depth map or disparity values between these points.

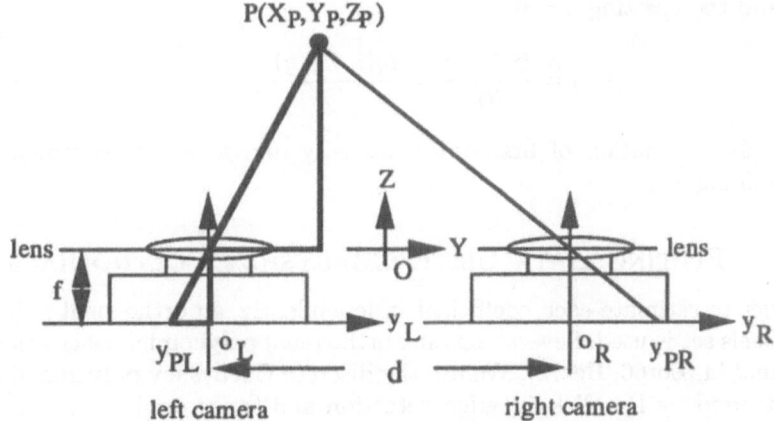

FIGURE 3.2. Camera geometry for stereo photography.

3.3   Estimation of Intensity Derivatives

Natural digital images are usually corrupted by a certain amount of noise due to the electronic imaging sensor, film granularity and quantization error. The derivatives obtained using a difference operator applied to digital images are not reliable. Since a digital image is obtained by sampling an analog image, a proper way to recover a smooth and continuous image surface is by a polynomial fitting technique. We first assume that a point at the right image corresponding to a specified point in the left image lies somewhere on the corresponding epipolar line which is parallel to the row coordinate, thus in a horizontal direction. Second, we assume in each neighborhood of the image the underlying intensity function can be approximated by a fourth order polynomial. The first assumption is also known as the epipolar constraint. With the help of this constraint, the first order intensity derivatives we need for matching are computed only for the horizontal direction. Under the second assumption, the intensity function in a window, centered at the point (i, j), of size $2\omega + 1$ is represented by a polynomial of the form

$$g(i, j + y) = a_1 + a_2 y + a_3 y^2 + a_4 y^3 + a_5 y^4 \qquad (3.6)$$

where y is lies in the range $-\omega$ to $+\omega$ and $\{a_i\}$ are the coefficients. If the window size is 3, then a second order polynomial is sufficient to represent the intensity function. The first order intensity derivative at point (i, j) can be easily obtained by taking the derivative about $g(i, j + y)$ with respect to y and then setting $y = 0$

$$g'(i,j) \triangleq \frac{\partial g(i,j)}{\partial j} = \frac{dg(i, j+y)}{dy}\Big|_{y=0} = a_2. \qquad (3.7)$$

Thus, the estimation of first order intensity derivatives is equivalent to determining a_2.

3.3.1   FITTING DATA USING CHEBYSHEV POLYNOMIALS

In order to estimate each coefficient independently, an orthogonal polynomial basis set is used. Several existing orthogonal polynomial basis sets can be found in [Lor66, Bec73]. We use the discrete Chebyshev polynomial basis set, used by Haralick for edge detection and topographic classification [Har84, LHW82]. An important property of using polynomials is that low order fits over a large window can reduce the effects of noise and give a smooth function.

Let a set of discrete Chebyshev polynomials be defined over an index set $\Omega = \{-\omega, -\omega + 1, ..., \omega - 1, \omega\}$, over a window of size $2\omega + 1$, as

$$
\begin{aligned}
Ch_0(y) &= 1 \\
Ch_1(y) &= y \\
Ch_2(y) &= y^2 - q_2/q_0 \\
Ch_3(y) &= y^3 - (q_4/q_2)\,y \\
Ch_4(y) &= y^4 + [(q_2 q_4 - q_0 q_6)\,y^2 + (q_2 q_6 - q_4^2)]/(q_0 q_4 - q_2^2)
\end{aligned}
\qquad (3.8)
$$

where

$$q_n = \sum_{k \in \Omega} k^n.$$

With the window centered at point (i, j), the intensity function $g(i, j+y)$ for each $y \in \Omega$ can be obtained as

$$\hat{g}(i, j+y) = \sum_{m=0}^{4} d_m\, Ch_m(y) \qquad (3.9)$$

where $\hat{g}(i, j + y)$ denotes the approximated continuous intensity function. For $\omega = 1$, only the first three Chebyshev polynomials are needed. By minimizing the least square error in estimation and taking advantage of the orthogonality of the polynomial set, the coefficients $\{d_m\}$ are obtained as

$$d_m = \frac{\sum_{y \in \Omega} Ch_m(y)\, g(i, j+y)}{\sum_{u \in \Omega} Ch_m^2(u)} \qquad (3.10)$$

where $\{g(i, j + y)\}$ are the observed intensity values.

Expanding (3.9) and comparing with (3.6) results in the first order intensity derivative coefficient a_2

$$a_2 = d_1 - \frac{q_4}{q_2}d_3$$

$$= \sum_{y \in \Omega} M(y)\, g(i, j+y) \tag{3.11}$$

where $M(y)$ is determined by

$$M(y) = \frac{Ch_1(y)}{\sum_{u \in \Omega} Ch_1^2(u)} - \frac{q_4}{q_2} \frac{Ch_3(y)}{\sum_{u \in \Omega} Ch_3^2(u)}. \tag{3.12}$$

For $\omega = 1$, the second term in (3.12) is zero. From (3.11) one can see that $M(y)$ is a filter for detecting intensity changes.

3.3.2 ANALYSIS OF FILTER $M(y)$

Basically, the filter $M(y)$ used for detecting intensity changes has to satisfy the following requirements. First, it should eliminate the amplitude bias completely. Second, it should remove noise very efficiently.

For simplicity of notation, we rewrite (3.11) as

$$a_2 = M(j) * g(i, j) \tag{3.13}$$

where "$*$" denotes the convolution operator. Suppose that the image is corrupted by amplitude bias b and additive white noise $\{n_{i,j}\}$ with zero mean and variance σ_n^2. The observed image is

$$\tilde{g}(i, j) = g(i, j) + b + n(i, j) \tag{3.14}$$

where $\tilde{g}(i, j)$ and $g(i, j)$ are the corrupted and original intensity functions, respectively. Noting that the filter $M(j)$ is an anti-symmetric function of j, the amplitude bias b is completely eliminated after the convolution operation. Therefore,

$$M(j) * \tilde{g}(i, j) = M(j) * (g(i, j) + n(i, j)). \tag{3.15}$$

The expected value of the filter output can be written as

$$\mathbf{E}\{M(j) * \tilde{g}(i, j)\} = M(j) * g(i, j). \tag{3.16}$$

Accordingly, the variance can be expressed as

$$\begin{aligned}
&\mathbf{E}\{(M(j) * \tilde{g}(i, j) - \mathbf{E}\{M(j) * \tilde{g}(i, j)\})^2\} \\
&= \mathbf{E}\{(M(j) * n(i, j))^2\} \\
&= \sigma^2 \sum_{j \in \Omega} M^2(j).
\end{aligned} \tag{3.17}$$

By using (3.12), it is straightforward to prove that

$$\sum_{j \in \Omega} M^2(j) = \frac{q_6}{q_6 q_2 - q_4^2} \tag{3.18}$$

where

$$q_i = \sum_{y \in \Omega} y^i.$$

Hence, the variance of the filter output is

$$\mathbf{E}\{(M(j) * \tilde{g}(i,j) - \mathbf{E}\{M(j) * \tilde{g}(i,j)\})^2\} = \frac{\sigma_n^2 q_6}{q_6 q_2 - q_4^2}. \tag{3.19}$$

For large window size, $q_6 \gg q_2$. The variance can be approximated as

$$\mathbf{E}\{(M(j) * \tilde{g}(i,j) - \mathbf{E}\{M(j) * \tilde{g}(i,j)\})^2\} = \frac{\sigma_n^2}{q_2}. \tag{3.20}$$

From (3.20), one can see that the variance becomes smaller and smaller as the window size increases. For instance, if the window size is 5, then the variance is $0.9\sigma^2$. If the window size is 11, then the variance is significantly reduced to $0.009\sigma^2$. However, a large window causes some loss of local information due to smoothing which smears or erases local features. If one desires to retain local features, then a small window may be used, but more noise remains and the estimated intensity function is rough. Also in order to reduce the effect of spatial quantization error for the natural images, a window as small as size 3 may be used, as discussed in the next section. The variance of the estimated derivatives using a 3×3 window is the same as that in (3.20). It appears that the choice of the window size is closely related to the theory of the human visual system. It is known [WG77, WB79] that at least four different size channels exist in a human visual system. Marr suggested [Mar82] that in order to detect intensity changes efficiently, the filter used should first be a differential operator, taking either a first or second order spatial derivative of the image, and second be capable of being tuned to act at any appropriate scale.

The following examples show that by choosing a proper window size, the effects of noise can be very efficiently eliminated. A 256×256 real image is used in these examples.

Example 1: An amplitude bias of strength 20 and white Gaussian noise (30 dB SNR) were added to the image. A section of the image is shown in Figure 3.3. The dashed and solid lines in Figure 3.3(a) represent the original and noisy images, respectively. Obviously, there is no way to match these two images based on the noisy biased intensity values alone. Figure 3.3(b) shows the estimated first order intensity derivatives from these two images using the polynomial method. The window size is 5, so the index set is $\{-2, -1, 0, 1, 2\}$.

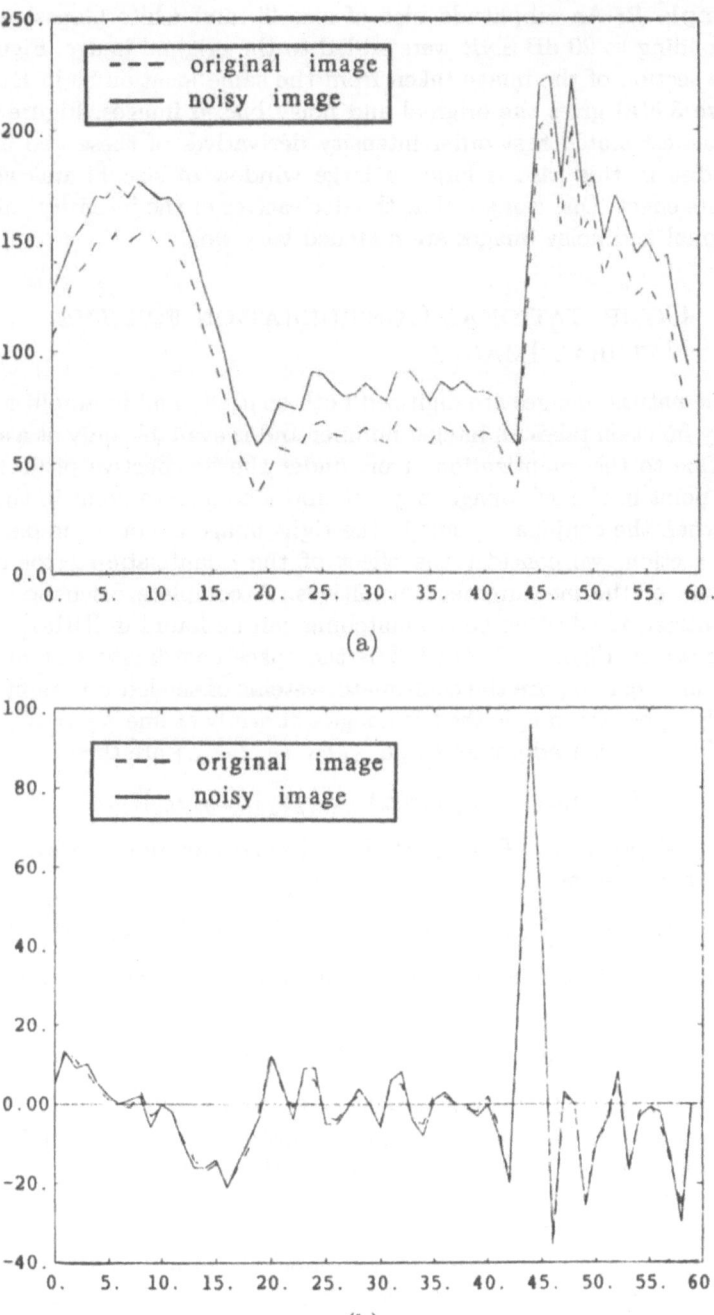

FIGURE 3.3. A section of a real image with amplitude bias 20 and 30 dB noise. (a) Intensity values of original and noisy images. (b) First order derivatives of intensity values of original and noisy images.

Example 2: An amplitude bias of size 20 and white Gaussian noise corresponding to 20 dB SNR were added to the original image. Figure 3.4 shows a section of the image taken from the same location as in Example 1. Figure 3.3(a) gives the original and noisy biased images. Figure 3.3(b) shows the estimated first order intensity derivatives of these two images. Since noise in this case is large, a large window of size 11 was used to reduce its effect. One can see that the derivatives of the intensity values of the original and noisy images are matched very well.

3.3.3   COMPUTATIONAL CONSIDERATION FOR THE NATURAL IMAGES

Since the natural images are digitized both spatially and in amplitude, the intensity function takes an integer number and is available only at a sample point. Due to the quantization error, under the perspective projection, a sample point in the left image may not find a conjugate point in the right image when the conjugate point in the right image is not a sample point. In this section, we consider the effect of the quantization error on the estimation of the measurement primitives. A complete discussion about the quantization effect on stereo matching can be found in [BH87].

As shown in Figure 3.2, $OXYZ$ is the space coordinate system, while $o_L x_L y_L$ and $o_R x_R y_R$ are the coordinate systems of the left and right image planes. Suppose we sample the left image uniformly at line $X_L = X_R = X_0$. A set of equally spaced points $\{..., L_{-1}, L_0, L_1, L_2, ...\}$ are then obtained at

$$\{..., (x_0, y_{L_{-1}}), (x_0, y_{L_0}), (x_0, y_{L_1}), (x_0, y_{L_2}), ...\}.$$

The original points $\{..., P_{-1}, P_0, P_1, P_2, ...\}$ corresponding to these sample points are located at

$$\{..., (X_0, Y_{-1}, Z_{-1}), (X_0, Y_0, Z_0), (X_0, Y_1, Z_1), (X_0, Y_2, Z_2), ...\}$$

on the object surface. These original points also project into the right image plane at

$$\{..., (x_0, y_{R_{-1}}), (x_{R_0}, y_{R_0}), (x_0, y_{R_1}), (x_0, y_{R_2}), ...\}.$$

When the object surface is not parallel to the image plane, the original object points are not equally-spaced on the surface. The corresponding right image points are not equally-spaced either. Hence, not all the corresponding right image points are sample points. This phenomenon is shown in Figure 3.5.

Let us assume that in the left image plane, the sample points match the image points exactly, and in the right image plane, only the image point R_0 matches the sample point (as illustrated in Figure 3.5) and the other image points do not match the sample points. Then we have

$$y_{L_i} - y_{L_0} = y_{L_i^s} - y_{L_0^s} \tag{3.21}$$

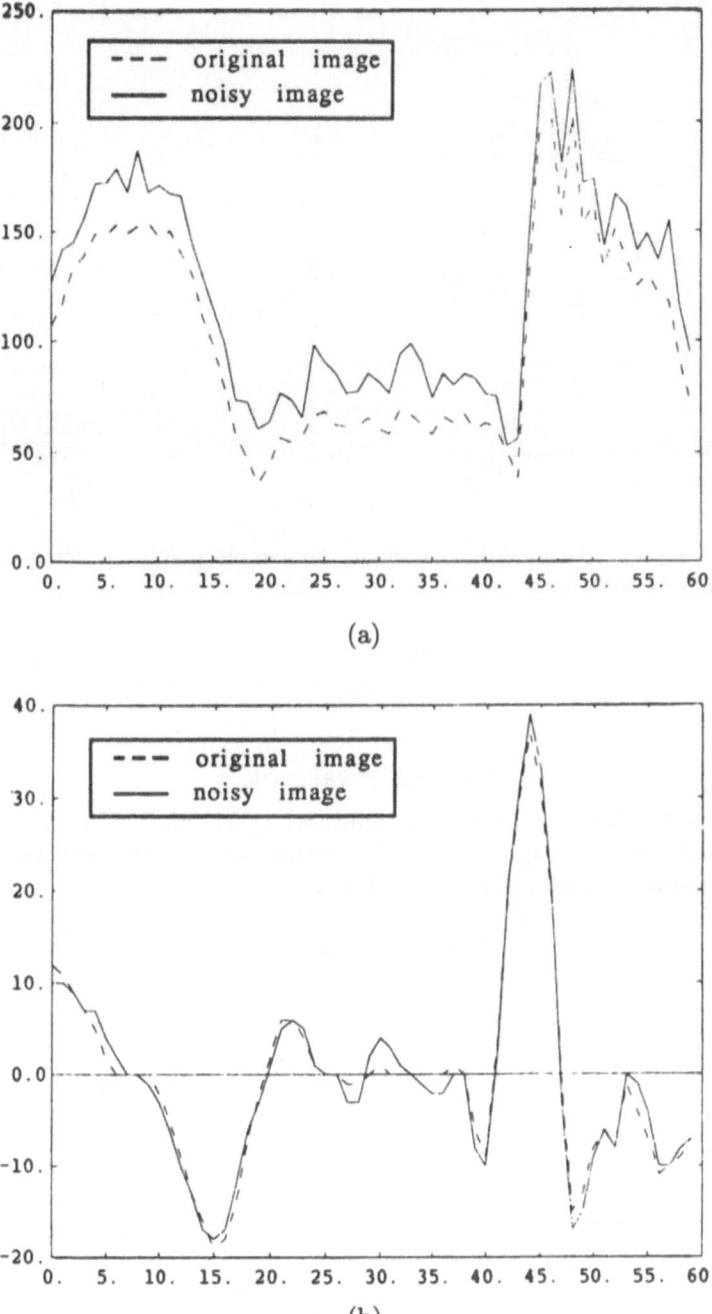

FIGURE 3.4. A section of a real image with amplitude bias 20 and 20 dB noise. (a) Intensity values of original and noisy images. (b) First order derivatives of intensity values of original and noisy images.

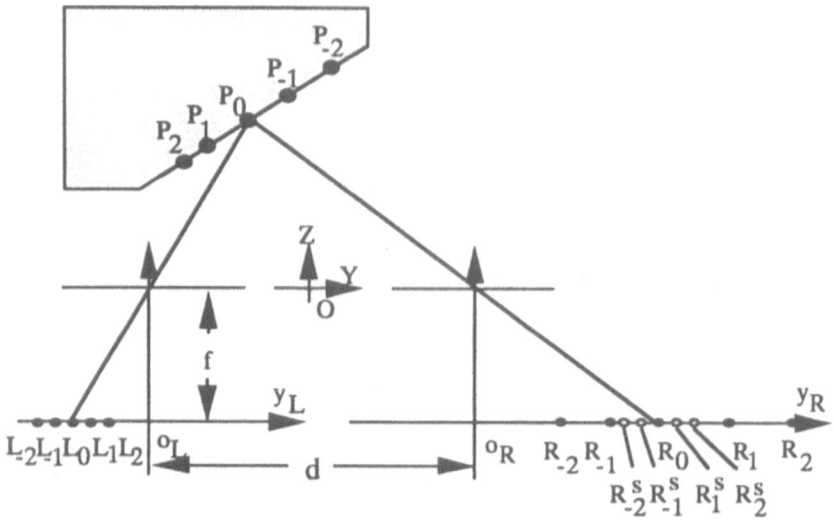

FIGURE 3.5. On the right image plane, the sample points do not match the image points everywhere.

and
$$y_{L_i} - y_{L_0} = y_{R_i^s} - y_{R_0} \qquad (3.22)$$

where the "s" denotes the sample point. By (3.21) and (3.22), the spatial quantization error, the distance between the sample point, and the corresponding image point can be computed as

$$\eta_i = \begin{cases} (y_{R_i} - y_{R_0}) - (y_{R_i^s} - y_{R_0}) = f\,d\,(\frac{1}{Z_0} - \frac{1}{Z_i}), & i > 0 \\ (y_{R_0} - y_{R_i}) - (y_{R_0} - y_{R_i^s}) = f\,d\,(\frac{1}{Z_i} - \frac{1}{Z_0}), & i < 0. \end{cases} \qquad (3.23)$$

Obviously,
$$\eta_i > \eta_{i-1}, \quad i > 0,$$

and
$$\eta_i < \eta_{i-1}, \quad i < 0.$$

This shows that the spatial quantization error depends on the depth Z, the focal length f and the distance d between the cameras. If the object surface is parallel to the image plane, then

$$Z_0 = Z_i, \quad \forall\ i$$

leading to

$$\eta_i = 0, \quad \forall\ i,$$

which means that the sample points match the corresponding image points perfectly.

An interesting aspect of (3.23) is that by definition of disparity in (3.5) the spatial quantization error is equal to the difference of the disparities between the points P_0 and P_i. Therefore, stereo matching algorithms using intensity values as the measurement primitives cannot detect such a difference if the sample interval is twice as large as the spatial quantization error.

We assume that the incident illumination and absorption characteristics of the object surface are roughly constant, and the surface orientation and the distance to the two cameras are almost the same. Therefore, the left and right image planes receive the same amount of light, which means that the intensity functions of conjugate image points are almost the same. Assuming that the intensity function $g(x_0, y_{R_i^s})$ is differentiable with respect to y, it can be expanded by a Taylor series about $(x_0, y_{R_i^s}) = (x_0, y_{R_i})$ as

$$g(x_0, y_{R_i^s}) = \begin{cases} g(x_0, y_{R_i} - \eta_i), & i > 0 \\ g(x_0, y_{R_i} + \eta_i), & i < 0 \end{cases} \tag{3.24}$$

where

$$g(x_0, y_{R_i} - \eta_i) = g(x_0, y_{R_i}) - \eta_i\, g'(x_0, y_R)|_{y_R = y_{R_i}} + O(\eta_i^2)$$

$$g(x_0, y_{R_i} + \eta_i) = g(x_0, y_{R_i}) + \eta_i\, g'(x_0, y_R)|_{y_R = y_{R_i}} + O(\eta_i^2)$$

and $g'(x_0, y_R)|_{y_R = y_{R_i}}$ represents the derivative of the intensity function at (x_0, y_{R_i}). By using the sampled intensity function to estimate the first order derivative of the intensity function $g(x_0, y_{R_0})$, (3.11) becomes

$$\tilde{g}'(x_0, y_{R_0}) = \sum_{y \in \Omega} M(y)\, g(x_0, (y_{R_0} + y)^s) \tag{3.25}$$

where the "~" denotes the estimated intensity derivative using the sampled intensity functions.

Replacing the sampled intensity functions in (3.25) by (3.24), we have

$$\tilde{g}'(x_0, y_R)|_{y_R = y_{R_0}} \simeq \sum_{i \in \Omega} M(i)\, g(x_{R_0}, y_{R_i})$$

$$+ \sum_{i \in \Omega} u(i)\, \eta_i\, M(i)\, g'(x_0, y_R)|_{y_R = y_{R_i}} \tag{3.26}$$

where $u(i)$ is a step function

$$u(i) = \begin{cases} 1, & i > 0 \\ -1, & i < 0. \end{cases}$$

Clearly, the first term in the right side of (3.26) is equal to (3.11), which means it is a correct estimate, while the second term is an estimation error caused by the spatial quantization error. Since the spatial quantization error is proportional to f and d, and is inversely proportional to Z, the estimation error is small when the camera is close enough and/or the object is far enough. When the surface is not parallel to the image plane and the object is close to the camera, using a large window to estimate the derivatives will give a large error due to the accumulated quantization error. Hence, a small window is preferred if the object is close to the camera. As proposed in [MPH79], the smallest channel in the human visual system contains a filter with a central diameter of $1.5'$, roughly corresponding to 4 pixels. Therefore, considering the effects of noise distortion and the spatial quantization error, a filter $m(y)$ the size of $3 - 7$ pixels is appropriate for natural stereo images.

In fact, (3.26) can be considered for both derivative estimation and edge detection since most edge detection algorithms can be considered as window operations followed by appropriate thresholding. Owing to the output error of the window operation, the edge detector may miss an edge, give a false edge or shift the edge. In other words, the edge output also suffers from the spatial quantization error.

Noting that the filter $m(y)$ is an anti-symmetric function of y and assuming that the derivatives at sample points are the same, (3.26) can be simplified as

$$\tilde{g}'(x_0, y_r)|_{y_r=y_{r_0}} \simeq g'(x_0, y_r)|_{y_r=y_{r_0}} \left[1 - \sum_{i=1}^{\omega} m(i) \left(\eta_i + \eta_{-i}\right)\right]. \qquad (3.27)$$

Substituting (3.23) and then (3.5), we finally have

$$\tilde{g}'(x_0, y_r)|_{y_r=y_{r_0}} \simeq g'(x_0, y_r)|_{y_r=y_{r_0}} \left[1 - \sum_{i=1}^{\omega} m(i) \left(D_i - D_{-i}\right)\right]. \qquad (3.28)$$

The estimate of the derivatives may be either larger or smaller than the true value depending on the orientation of the object surface.

3.4 Matching Using a Network

Binary neurons are used to represent the disparity values between two images. The network consists of $N_r \times N_c \times (D+1)$ mutually interconnected neurons, where D is the maximum disparity, and N_r and N_c are the image row and column sizes, respectively. Let $V = \{v_{i,j,k}, 1 \le i \le N_r, 1 \le j \le N_c, 0 \le k \le D\}$ be a binary state set of the neural network with $v_{i,j,k}$ (1 for firing and 0 for resting) denoting the state of the (i, j, k)th neuron. When $v_{i,j,k}$ is 1, this means that the disparity value is k at the point (i, j). Every point is represented by $D+1$ mutually exclusive neurons, signifying that only one neuron is firing while the others are resting, as a result of the uniqueness constraint of the matching problem.

The network parameters, the interconnection strengths $T_{i,j,k;l,m,n}$ and the bias inputs $I_{i,j,k}$, can be determined in terms of the energy function of the network. As defined by (2.7), the energy function of the neural network can be written as

$$E = -\frac{1}{2} \sum_{i=1}^{N_r} \sum_{l=1}^{N_r} \sum_{j=1}^{N_c} \sum_{m=1}^{N_c} \sum_{k=0}^{D} \sum_{n=0}^{D} T_{i,j,k;l,m,n}\, v_{i,j,k}\, v_{l,m,n}$$

$$- \sum_{i=1}^{N_r} \sum_{j=1}^{N_c} \sum_{k=0}^{D} I_{i,j,k}\, v_{i,j,k}. \qquad (3.29)$$

In order to use the spontaneous energy–minimization process of the neural network, we reformulate the stereo matching problem under the epipolar assumption as one of minimizing an error function with constraints defined as

$$E = \sum_{i=1}^{N_r} \sum_{j=1}^{N_c} \sum_{k=0}^{D} (g_l'(i,j) - g_r'(i,j+k))^2\, v_{i,j,k}$$

$$+ \frac{\lambda}{2} \sum_{i=1}^{N_r} \sum_{j=1}^{N_c} \sum_{k=0}^{D} \sum_{s \in S} (v_{i,j,k} - v_{(i,j)+s,k})^2 \qquad (3.30)$$

where $\{g_L'(\cdot)\}$ and $\{g_R'(\cdot)\}$ are the first order intensity derivatives of the left and right images, respectively, S is an index set excluding $(0,0)$ for all the neighbors in a $\Gamma \times \Gamma$ window centered at point (i,j), and λ is a constant. The first term in (3.30) is called the photometric constraint, which seeks disparity values such that all regions of two images are matched in a least squares sense. Meanwhile, the second term is the smoothness constraint on the solution. The constant λ determines the tradeoff between the two terms to achieve the best results.

Expanding (3.30), we get

$$E = \sum_{i=1}^{N_r} \sum_{j=1}^{N_c} \sum_{k=0}^{D} (g_l'(i,j) - g_r'(i,j+k))^2\, v_{i,j,k}$$

$$+ \frac{\lambda}{2} \sum_{i=1}^{N_r} \sum_{j=1}^{N_c} \sum_{k=0}^{D} [W\, v_{i,j,k}^2$$

$$- \sum_{s \in S} (2 v_{i,j,k} v_{(i,j)+s,k} - v_{(i,j)+s,k}^2)] \qquad (3.31)$$

where W is the weight depending on the window size Γ. If the boundary conditions are ignored, then $v_{i,j,k}^2$ and $v_{(i,j)+s,k}^2$ can be combined into one term

$$E^- = \sum_{i=1}^{N_r} \sum_{j=1}^{N_c} \sum_{k=0}^{D} (g_l'(i,j) - g_r'(i,j+k))^2\, v_{i,j,k}$$

$$+\frac{\lambda}{2} \sum_{i=1}^{N_r} \sum_{j=1}^{N_c} \sum_{k=0}^{D} 2\left(W v_{i,j,k}^2 - \sum_{s\in S} v_{i,j,k} v_{(i,j)+s,k}\right). \quad (3.32)$$

Comparing the terms in the expansion of (3.32) with the corresponding terms in (3.29), we can determine the interconnection strengths and bias inputs as

$$T_{i,j,k;l,m,n} = -2W\lambda \delta_{i,l}\delta_{j,m}\delta_{k,n} + 2\lambda \sum_{s\in S} \delta_{(i,j),(l,m)+s}\delta_{k,n} \quad (3.33)$$

and

$$I_{i,j,k} = -(g_L'(i,j) - g_R'(i,j+k))^2 \quad (3.34)$$

where $\delta_{a,b}$ is the Dirac delta function. The weight W is determined by

$$W = \Gamma^2 - 1.$$

From (3.33) it can be seen that the interconnections are symmetric and the self-connection $T_{i,j,k;i,j,k}$ is not zero. The non-zero self-connection requires self-feedback for neurons.

Matching is carried out by neuron evaluation. Once the parameters, interconnection strengths $T_{i,j,k;l,m,n}$ and and bias inputs $I_{i,j,k}$ are obtained using (3.33) and (3.34), each neuron can randomly and synchronously (or asynchronously) evaluate its state and readjust according to

$$u_{i,j,k} = \sum_{l=1}^{N_r} \sum_{m=1}^{N_c} \sum_{n=0}^{D} T_{i,j,k;l,m,n} v_{l,m,n} + I_{i,j,k} \quad (3.35)$$

and

$$v_{i,j,k} = g(u_{i,j,k}) \quad (3.36)$$

where $g(x_{i,j,k})$ is a maximum evolution function

$$g(x_{i,j,k}) = \begin{cases} 1 & if \ x_{i,j,k} = max(x_{i,j,l}; l = 0, 1, ..., D) \\ 0 & otherwise. \end{cases} \quad (3.37)$$

The synchronous updating scheme can be implemented in parallel, while the asynchronous updating scheme can be sequentially implemented. Another updating scheme called the hybrid updating scheme results when some neurons are synchronously updated and the others are asynchronously updated. For natural stereo images, we will use the hybrid neural network. The problem of uniqueness in matching is ensured by a batch updating scheme: $D + 1$ neurons $\{v_{i,j,0}, ...v_{i,j,D}\}$ at location (i,j) are updated at each step simultaneously.

The initial state of the neurons was set as

$$\cdot v_{i,j,k} = \begin{cases} 1 & if \ I_{i,j,k} = max(I_{i,j,l}; l = 0, 1, ..., D). \\ 0 & otherwise \end{cases} \quad (3.38)$$

where $I_{i,j,k}$ is the bias input.

As mentioned in Chapter 2, the self–feedback may cause the energy function to increase with a transition. The batch updating scheme simultaneously updates $(D+1)$ neurons $\{v_{i,j,k}; k = 0, ..., D\}$ corresponding to the image point (i, j) at each step. However, at most two of the $(D+1)$ neurons change their state at each step. By defining the state changes $\Delta v_{i,j,k}$ and $\Delta v_{i,j,k'}$ of neurons (i, j, k) and (i, j, k') and energy change ΔE as

$$\Delta v_{i,j,k} = v_{i,j,k}^{new} - v_{i,j,k}^{old}$$

$$\Delta v_{i,j,k'} = v_{i,j,k'}^{new} - v_{i,j,k'}^{old}$$

and

$$\Delta E = E^{new} - E^{old}$$

the change ΔE due to changes $\Delta v_{i,j,k}$ and $\Delta v_{i,j,k'}$ can be obtained as

$$
\begin{aligned}
\Delta E \;=\; & -\Big(\sum_{l=1}^{N_r}\sum_{m=1}^{N_c}\sum_{n=0}^{D} T_{i,j,k;l,m,n}\, v_{l,m,n} + I_{i,j,k}\Big)\,\Delta v_{i,j,k} \\
& -\Big(\sum_{l=1}^{N_r}\sum_{m=1}^{N_c}\sum_{n=0}^{D} T_{i,j,k';l,m,n}\, v_{l,m,n} + I_{i,j,k'}\Big)\,\Delta v_{i,j,k'} \\
& -\frac{1}{2}\,T_{i,j,k;i,j,k}\,(\Delta v_{i,j,k})^2 - \frac{1}{2}\,T_{i,j,k';i,j,k'}\,(\Delta v_{i,j,k'})^2 \\
& -T_{i,j,k;i,j,k'}(\Delta v_{i,j,k}v_{i,j,k'}^{new} + \Delta v_{i,j,k'}v_{i,j,k}^{new}).
\end{aligned}
\tag{3.39}
$$

When

$$v_{i,j,k}^{old} = 0, \qquad v_{i,j,k'}^{old} = 1,$$

$$u_{i,j,k} > u_{i,j,k'}$$

and the maximum evolution function is as in (3.37), we have

$$v_{i,j,k}^{new} = 1, \qquad v_{i,j,k'}^{new} = 0$$

and

$$\Delta v_{i,j,k} = 1, \qquad \Delta v_{i,j,k'} = -1.$$

Noting that

$$T_{i,j,k;i,j,k'} = 0 \quad if \quad k \neq k',$$

(3.39) can be simplified as

$$\Delta E = (u_{i,j,k'} - u_{i,j,k}) - \frac{1}{2}\,(T_{i,j,k;i,j,k} + T_{i,j,k';i,j,k'}).\tag{3.40}$$

Thus, the first term in (3.40) is negative. But

$$T_{i,j,k;i,j,k} + T_{i,j,k';i,j,k'} = -96\,\lambda < 0$$

leading to
$$-\frac{1}{2}\left(T_{i,j,k;i,j,k}+T_{i,j,k';i,j,k'}\right)>0.$$

When the first term is less than the second term in (3.39), then $\Delta E>0$.

A deterministic decision rule is used to ensure convergence of the network, probably to a local minimum. The stereo matching algorithm can then be summarized as

1. Set the initial state of the neurons.

2. Update the state of all neurons randomly and asynchronously (or synchronously) according to the deterministic decision rule.

3. Check the energy function; if the energy does not change anymore, stop; otherwise, go back to Step 2.

3.5 Experimental Results

A variety of images including random dot stereograms and natural stereo image pairs were tested using our algorithm. A 5×5 ($\Gamma=5$) smoothing window was used for all images.

3.5.1 RANDOM DOT STEREOGRAMS

The random dot stereograms were created by the pseudo random number generating method described in [Jul60]. Each dot consists of only one element. All the following random dot stereograms are of size 128×128 and in the form of a three level "wedding cake". The background plane has zero disparity and each successive layer plane has an additional two elements of disparity. In order to implement this algorithm more efficiently on a conventional computer, we make the following simplifications. Since only one of $D+1$ neurons is firing at each point, we used one neuron lying in the range 0 to D to represent the disparity value instead of $D+1$ neurons. From (3.33) one can see that the interconnections between the neurons are local (a $\Gamma\times\Gamma$ neighborhood) and have the same structure for all neurons. Therefore, for $\Gamma=5$ we used a 5×5 window for computing $U_{i,j,k}$ and energy function E instead of a $N_rN_c(D+1)\times N_rN_c(D+1)$ interconnection strength matrix. The simplified algorithm greatly reduces the space complexity by increasing the program complexity a little. Therefore, it is very fast and efficient.

Figure 3.6 shows a 10% random dot stereogram. About 10% of the dots are white and the rest are black. Intensity values of the white and black elements are 255 and 0, respectively. Figure 3.6(a) is the left image and Figure 3.6(b) is the right image. Figure 3.6(c) is the resulting disparity map after 10 iterations using the asynchronous updating scheme. When the synchronous updating scheme is used, 23 iterations are needed. The disparity values are encoded as intensity values with the brightest value denoting the maximum disparity value. We used $\lambda = 20$, $D = 6$ and $\omega = 2$ (thus the window size was 5). Note that the disparity map is dense.

A similar test was run on the decorrelated stereogram [Jul71]. The original stereogram is 50% density random dots. In the left image, 20% of the dots were decorrelated at random as shown in Figure 3.7(a). By setting $\lambda = 2800$, $D = 6$ and $\omega = 2$, a dense disparity map in Figure 3.7(c) was obtained after 12 asynchronous iterations. The same result can be obtained after 19 synchronous iterations.

Another type of perturbation is Gaussian white noise [Jul60]. Figure 3.8 shows a pair of multiple gray level random dot images with intensity value in the range of $[0, 255]$. A 5 dB signal to noise ratio (SNR) Gaussian white noise was added to the left image. The SNR is defined as

$$SNR = 10 \ \log_{10} \ \frac{\sigma_o^2}{\sigma_n^2}$$

where σ_o^2 and σ_n^2 are the variances of the original left image and noise. The parameters $\lambda = 450$, $D = 6$ and $\omega = 2$ were used. Only six asynchronous iterations were needed to get the final result in Figure 3.8(c). When the synchronous updating scheme was used, nine iterations were needed to get the same result.

As expected, both the synchronous and asynchronous updating schemes work very well, although the latter takes more iterations. The synchronous updating scheme is suitable for parallel processing.

3.5.2 NATURAL STEREO IMAGES

Two stereo pairs of natural images, the Renault part and the Pentagon images, were used to test our algorithm. All images are of size 256×256. Since natural stereo images may not satisfy the epipolar constraint, small alignment corrections in the vertical direction are needed. A hybrid updating scheme was used for both the Renault and the Pentagon image pairs. The image is segmented into homogeneous and inhomogeneous regions by using a local variance criterion. A homogeneous region is defined as a smooth region with small local variances. The neurons corresponding to homogeneous image regions are updated sequentially, while the neurons corresponding to inhomogeneous regions are updated in parallel. Since the first derivatives of the intensity function in homogeneous regions are small, the inputs are small and the neurons tend to take the same state as their neighbors because of the smoothness constraint. No doubt, the neurons near the boundary will be first affected by the neighbors corresponding to inhomogeneous regions. As the neurons corresponding to homogeneous regions are-sequentially updated, they will all be affected by the boundary conditions, thus surfaces in homogeneous regions can be interpolated.

For the Renault images, the parameters were set as $\lambda = 12$, $D = 13$ and $\omega = 1$. The threshold for the local variances was set at 1.0. The local variance was computed over a 5×5 window. Approximately 72 iterations were required. Since a discrete network was used, the disparities only take integer values. A simple smoothing technique was applied to the nonzero portions of the disparity surface. Figures 3.9(a) and (b) show the left and right Renault part images. The disparity map is given in Figure 3.9(c). The smoothed version of the disparity map is shown in Figure 3.9(d). The smooth operator is a 9×9 mean filter. Figures 3.10 and 3.11 give the three-dimensional plots of the unsmoothed and smoothed disparity surfaces corresponding to disparity maps in Figures 3.9(c) and (d).

Figures 3.12(a) and (b) show the left and right Pentagon images. By choosing parameters $\lambda = 10$, $D = 4$, $\omega = 1$ and the local variance threshold 0.01, a disparity map was generated after 51 iterations. A 5×5 window was used for computing the local variance. Figures 3.12(c) and (d) give the unsmoothed and smoothed disparity maps, respectively. A 13×13 mean filter was used for smoothing. The three-dimensional plots for the disparity maps are shown in Figures 3.13 and 3.14, respectively.

3.6   Discussion

This chapter has presented an approach for extracting depth from a pair of static stereo images. A discrete neural network is employed to iteratively minimize the error function and an estimate of the disparity values is obtained when the neural network converges to a minimum. The experimental results offer support for the hypothesis that the first order derivatives of the intensity function may be considered to be an appropriate measurement primitive for the stereo matching problem. No surface interpolation step is involved in this approach because of the dense first order derivatives of the intensity function used as measurement primitives.

In considering the experimental results, however, one point should be kept in mind. It concerns the occluding pixels. At the location of the occluding pixels, the disparity values are undetermined. Hence, we have to detect the occluding pixels and establish some rules to infer the depth information at such a location. This issue will be discussed in the next chapter.

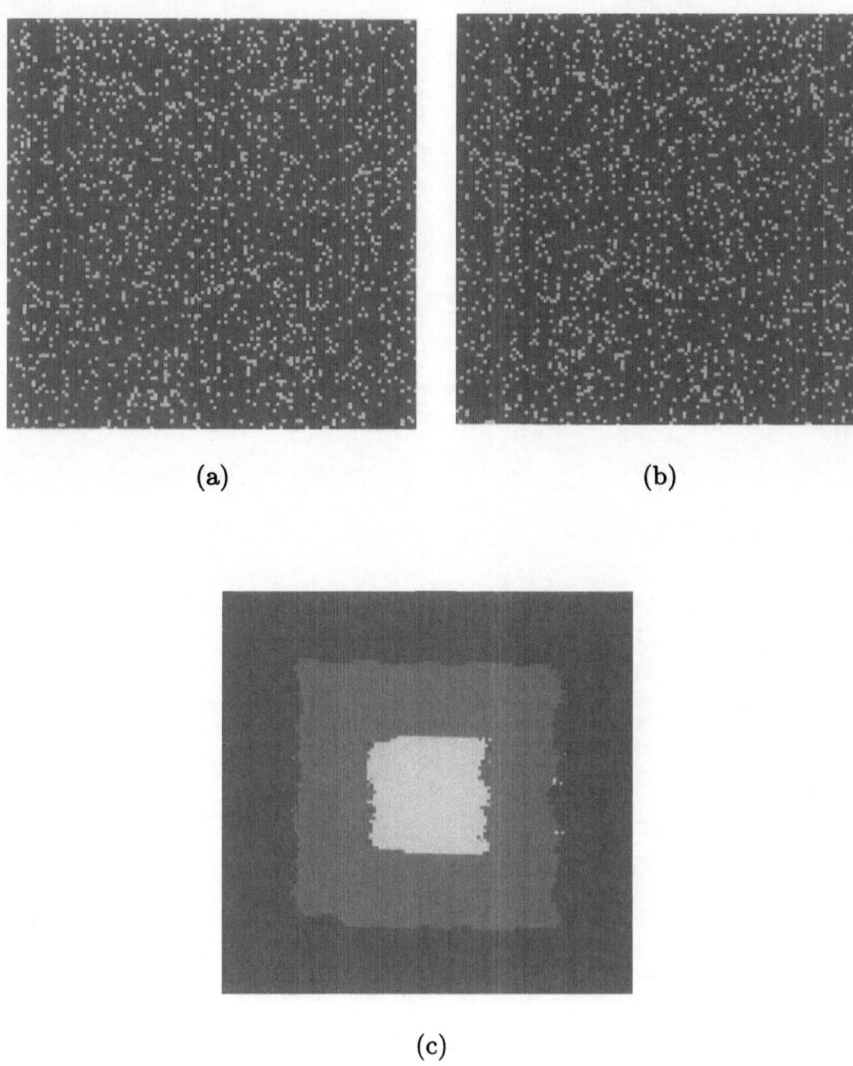

(a)   (b)

(c)

FIGURE 3.6. A 10% density random dot stereogram. (a) Left image. (b) Right image. (c) Test results. Disparity map represented by an intensity image. The intensity of the pixels corresponds to the height of the levels. (Reprinted with permission from [ZC88b] Figure 1, ©1988 IEEE.)

(a)

(b)

(c)

FIGURE 3.7. A 50% density random dot stereogram. (a) Left image, where 20% of the dots were decorrelated at random. (b) Right image. (c) Test results. The disparity map is represented by an intensity image. The intensity of the pixels corresponds to the height of the levels. (Reprinted with permission from [ZC88b] Figure 2, ©1988 IEEE.)

(a)   (b)

(c)

FIGURE 3.8. A multilevel random dot stereogram. (a) Left image, in which 5 dB SNR Gaussian white noise was added. (b) Right image. (c) Test Results. Disparity map represented by an intensity image. The intensity of the pixels corresponds to the height of the levels.

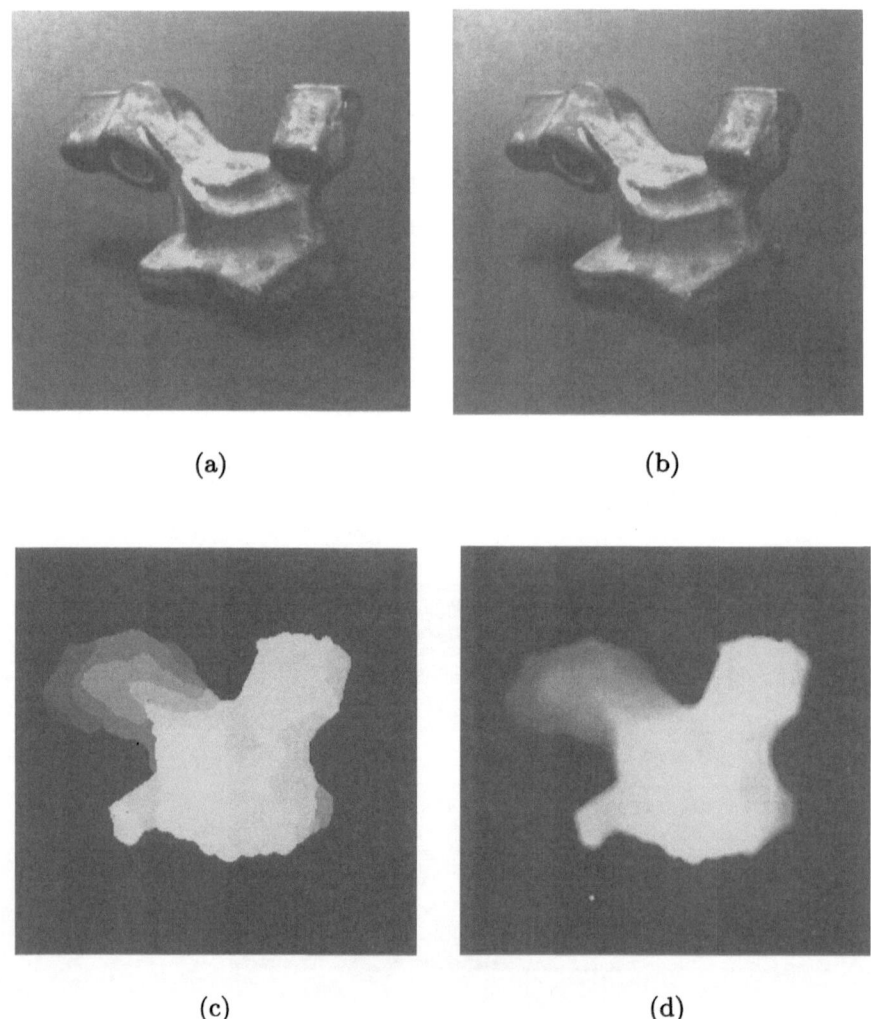

(a)   (b)

(c)   (d)

FIGURE 3.9. Renault part images. (a) Left image. (b) Right image. (c) Disparity map (reprinted from [ZC88a] Figure 3). (d) Smoothed disparity map.

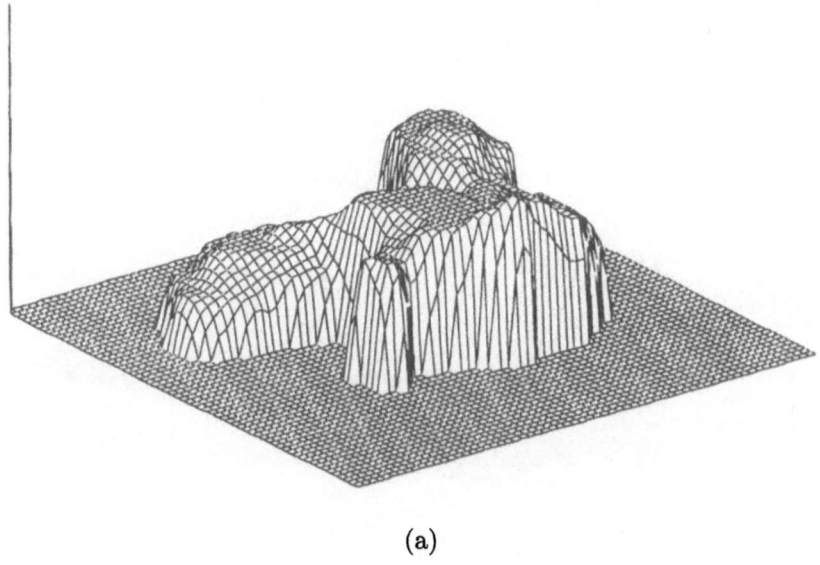

(a)

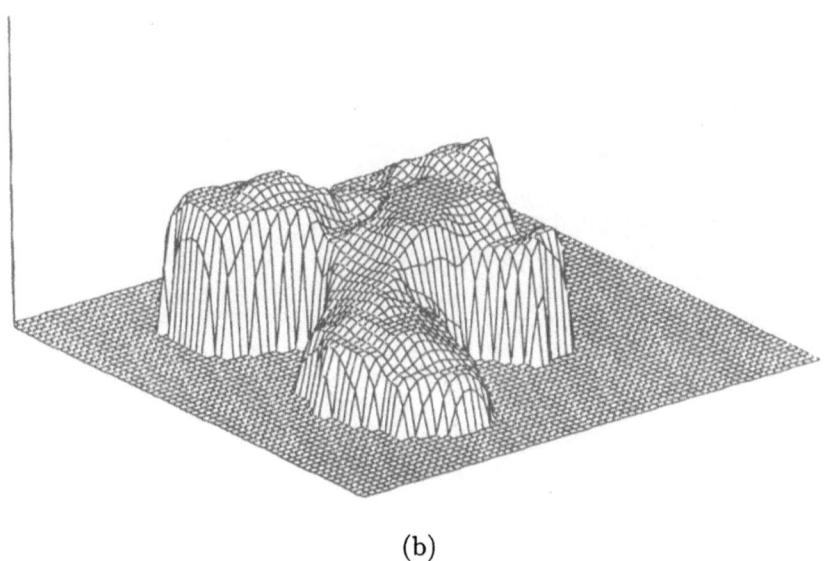

(b)

FIGURE 3.10. Three-dimensional plots of the disparity map for the Renault part images. (a) Front view. (b) Left side view. (Reprinted from [ZC88a] Figure 4.)

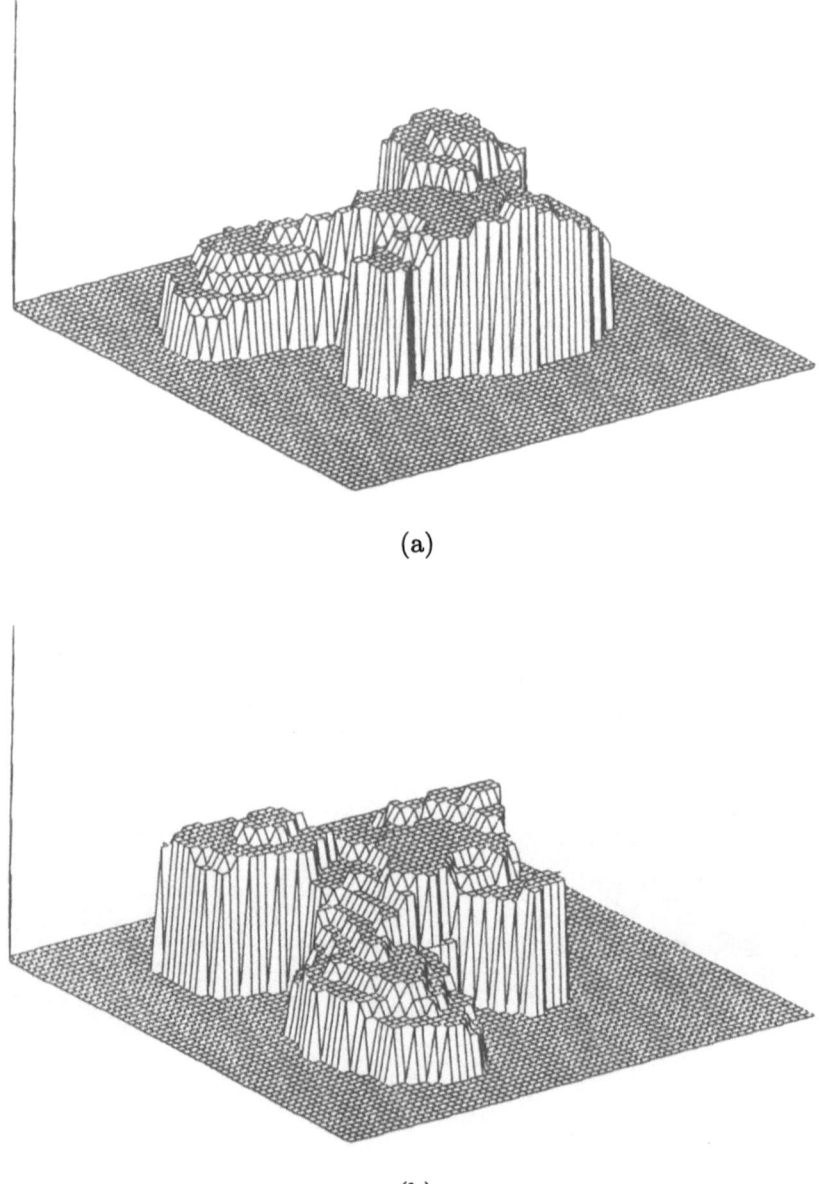

(a)

(b)

FIGURE 3.11. Three-dimensional plots for the smoothed disparity map of the Renault part images. (a) Front view. (b) Left side view.

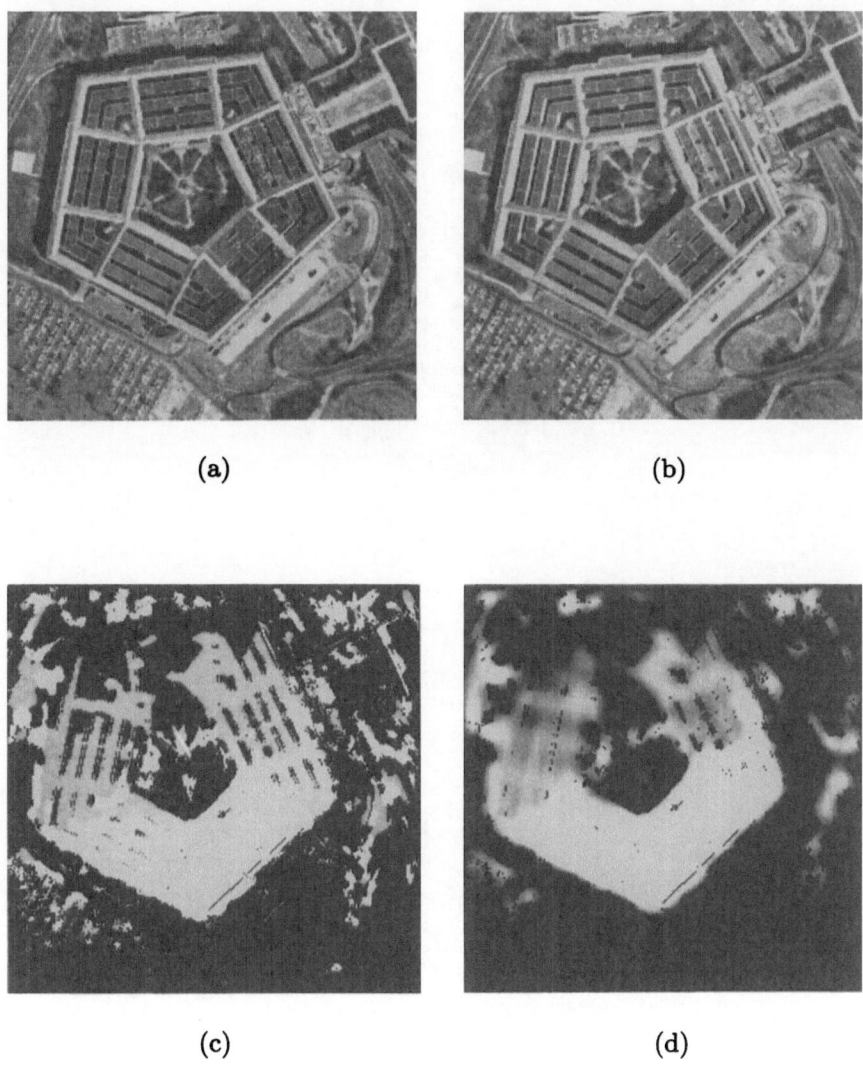

(a)   (b)

(c)   (d)

FIGURE 3.12. Pentagon images. (a) Left image. (b) Right image. (c) Disparity map. (d) Smoothed disparity map.

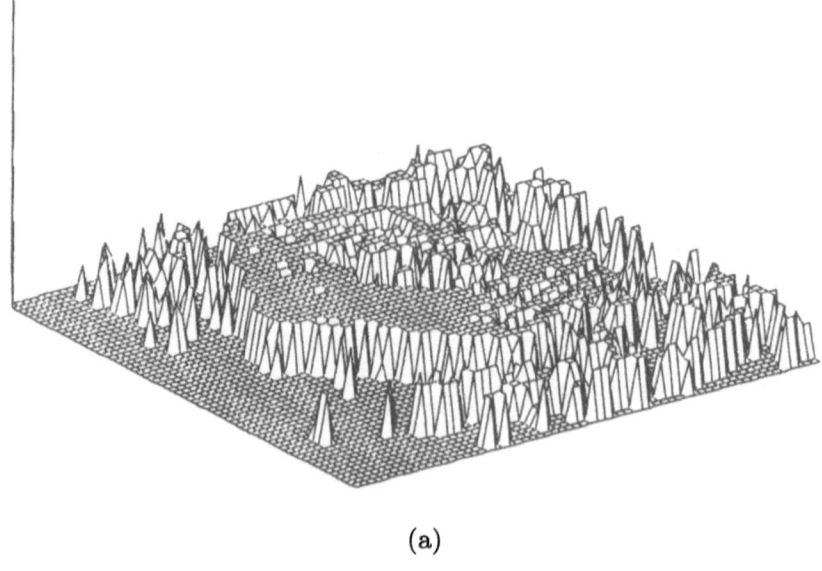

(a)

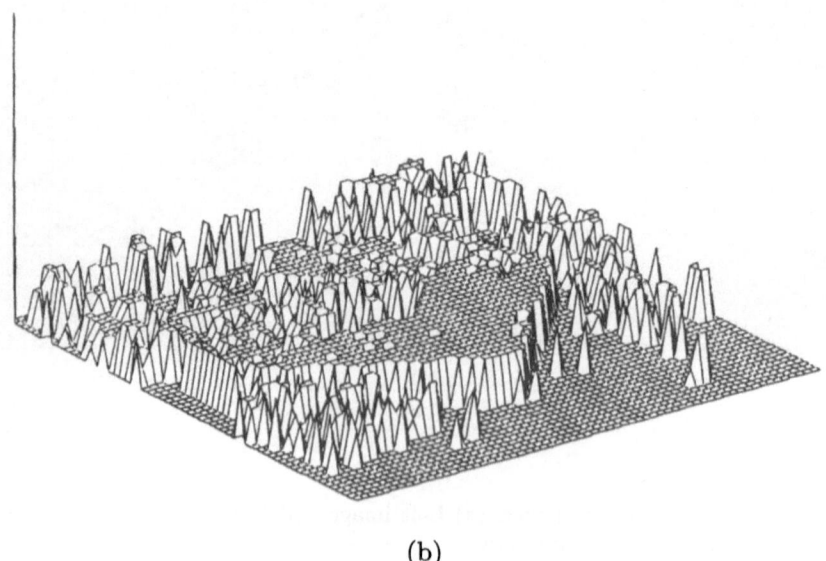

(b)

FIGURE 3.13. Three-dimensional plots for the disparity map of the Pentagon images. (a) Right side view. (b) Front view.

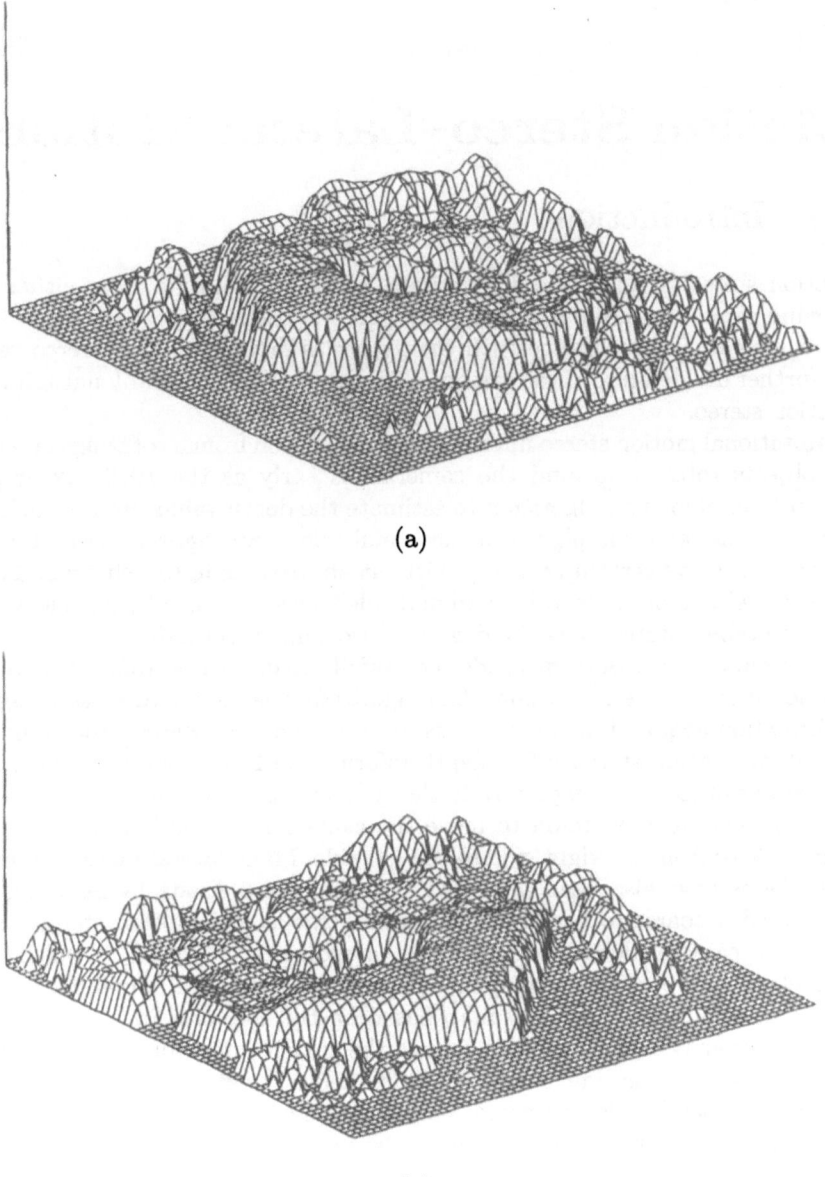

(a)

(b)

FIGURE 3.14. Three-dimensional plots for the smoothed disparity map of the Pentagon images. (a) Right side view. (b) Front view.

4

Motion Stereo–Lateral Motion

4.1 Introduction

Motion stereo is a method for deriving depth information from either a moving camera or objects moving through a stationary three-dimensional environment. In accordance with the nature of motion, motion stereo can be further divided into three categories: rotational, lateral and longitudinal motion stereo.

Rotational motion stereo infers depth information from a rotating camera or objects rotating around the camera. As early as the 1970s, Nevatia [Nev76] developed an algorithm to estimate the depth values from rotation motion. Basically, his algorithm uses multiple views between two stereo views to achieve certain accuracy without an increase in search time. The object is placed on a turntable and multiple views are taken by a camera as the turntable rotates every 0.5 degrees. Two simple methods are suggested for a region search. Both methods successfully reduce the search range, but do not increase the resolution. The algorithm does not make use of any information acquired in the previous view for the next search procedure.

Lateral motion stereo infers depth information from a laterally moving camera or objects moving laterally through a stationary three-dimensional environment. It is common to move the camera from the left side to the right side or from the right side to the left side. Many lateral motion stereo algorithms have also been proposed. Xu, Tsuji and Asada [XTA87] have suggested a coarse-to-fine iterative method for lateral motion stereo. By sliding a camera along a straight line, a sequence of images is taken at predetermined positions. The pair with the short baseline is matched first to produce a coarse disparity map based on the zero-crossings. Then the coarse disparity map is used to reduce the search range for the pair with the next longer baseline. This procedure is continued until the pair with the largest baseline is processed. One major advantage of this method is that occlusions can be predicted from the previous disparity map to avoid mismatches at the present step. Although the computation time is less compared to other coarse-to-fine methods, this method gives only a sparse disparity map and cannot be implemented on line.

Matthies, Szeliski and Kanade [MSK88] have introduced two real time approaches, based on intensity values and features using a Kalman filtering technique. A sequence of lateral motion images is generated by a moving camera along a straight line from left to right (or right to left). The intensity-based approach consists of four stages for each frame. First, a new measurement of disparity at each pixel is obtained by using a correlation matching procedure. Then the estimate of disparity is updated by a Kalman filter update equation based on the new measurement. Third, a generalized piecewise continuous spline technique is used to smooth the updated estimate. Finally, the disparity for each pixel in the next frame is given by the prediction procedure. As reported in [MSK88], the intensity-based approach is more efficient than the feature-based approach. However, a major problem in the intensity-based approach is that once the updated estimate is smoothed in the third stage, the gain of the Kalman filter and the error variance of the estimate are no longer correct, so they cannot be used in the next iteration.

Longitudinal motion stereo infers depth information from a forward moving camera or objects moving forward or backward through a stationary three-dimensional environment. Several algorithms have been developed relating to [Wil80] [BB85] Longitudinal motion stereo. Bolles and Baker [BB85]'s algorithms compute the depth values from the object path through the epipolar-plane images (EPIs). They assumed that the camera moves in a straight line laterally or forward. More details about their algorithms are given in the next chapter.

Rather than computing depth in image space, Jain, Bartlett and O'Brien [JBO87] developed a method for estimating the depth of feature points (corners) in the ego-motion complex logarithmic mapping (ECLM) space. They showed that the axial movement of the camera causes only horizontal but not vertical change in the mapping of image points. Therefore, the depth of a feature point can be determined from the horizontal displacement in the ECLM for that point and from the camera velocity in the gaze direction. However, the mapping is very sensitive to noise, spatial quantization error and image blur, requiring some heuristics to establish the correspondence of points, such as thresholds for maximum possible changes in the vertical direction and an upper bound for the search range in the horizontal direction in the ECLM space. Also the focus of expansion (FOE) for arbitrary translation of the camera and the feature points (corners) are assumed to be known. Another motion stereo method using feature points (corners) for computing depth in image space can be found in [IMO84].

In this chapter, we present two neural network-based algorithms, batch and recursive, for computing the disparity field using a sequence of lateral motion images. The batch approach first determines the bias inputs by using all image frames. A disparity field is then computed by the neural network. Since the batch approach computes the disparity field only once, computational complexity is reduced in comparison with existing approaches. The recursive approach uses a recursive least squares (RLS) algorithm to update the bias inputs instead of updating disparity values. When the next frame becomes available, the bias inputs are updated first and then a neural network is employed to estimate the disparity values under the epipolar, photometric and smoothness constraints. Unlike [MSK88],

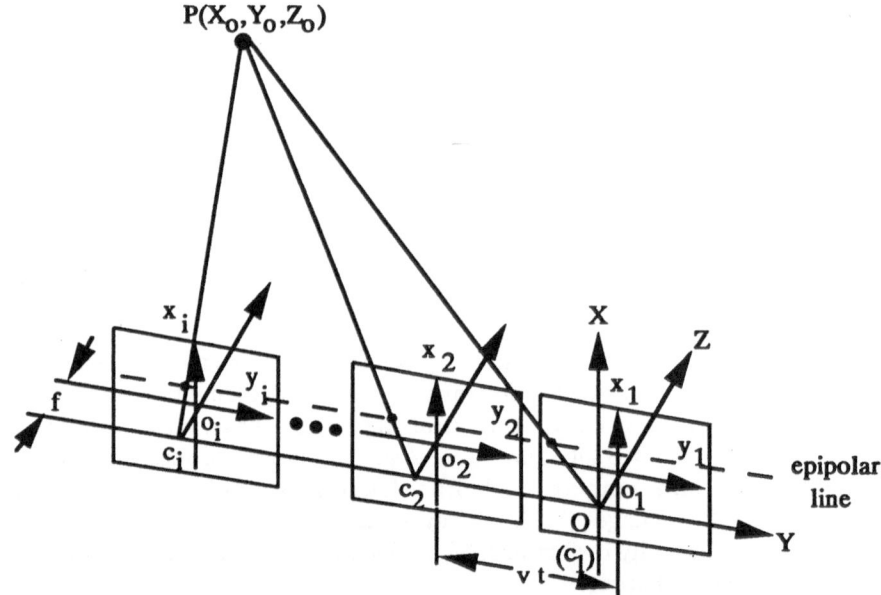

FIGURE 4.1. Camera geometry for motion stereo.

no smoothing procedure is needed. Since the neural network can be run in parallel and the RLS algorithm can be implemented on line, this approach is extremely fast and hence useful for real time robot vision applications. As the derivatives of the intensity function are more reliable than the intensity values and are also dense, both approaches use the derivatives as measurement primitives for matching. The Chamfer distance information is also used for matching to overcome the lack of information in homogeneous regions.

4.2 Depth from Lateral Motion

It is assumed that a sequence of images is taken by a camera moving with constant velocity from the right side to the left side along a straight line, as shown in Figure 4.1. For simplicity, we assume that the optical axis of the camera is perpendicular to the moving direction and the horizontal axes of the image planes are parallel to the moving direction. The constraint imposed on the camera configuration is to restrict the search within the horizontal direction, the so-called epipolar constraint. We also assume that the camera takes pictures exactly every t seconds apart such that all images are equally separated, so each successive image pair has the same baseline.

Let $OXYZ$ be the space coordinate system with the Z axis aligned with the camera optical axis and $o_i x_i y_i$ be the ith image plane coordinate system. The origin of the ith image system is located at $(0, -(i-1)vt, f)$ of the space system, where v is the velocity of camera, vt is the distance between two successive images and f is the focal length of the lens which takes a positive value in the space system. Under perspective projection, a point in space, (X_o, Y_o, Z_o), projects into the ith image plane at

$$(x_i, y_i) = (\frac{f\, X_0}{Z_o}, \frac{f\, (Y_o + (p-1)\, v\, t)}{Z_o}). \tag{4.1}$$

Theoretically, the disparity D_o can be derived from two successive image frames

$$D_o = y_p - y_{p-1} = \frac{f\, v\, t}{Z_o}, \tag{4.2}$$

which is the same as the formula used in static stereo. However, noise distortion, spatial quantization error and motion blur limit the estimation accuracy. In order to achieve high accuracy, a long sequence of image frames is required.

4.3   Estimation of Measurement Primitives

4.3.1   ESTIMATION OF DERIVATIVES

As only derivatives in the horizontal direction are required for matching, the epipolar constraint saves a lot of computations. By using a set of univariate discrete Chebyshev polynomials to approximate the intensity function within a window, we have

$$\hat{g}(i, j + y) = \mathbf{A}^t\, \mathbf{CH}(y) \tag{4.3}$$

where $\hat{g}(i, j + y)$ is the approximated continuous intensity function, t denotes the transpose operator,

$$\mathbf{A}^t = [a_0, a_1, a_2, a_3, a_4]$$

is the coefficient vector and

$$\mathbf{CH}^t(y) = [Ch_0(y),\ Ch_1(y),\ Ch_2(y),\ Ch_3(y),\ Ch_4(y)]$$

is the polynomial vector defined over an index set $\Omega = \{-\omega, -\omega+1, ..., \omega-1, \omega\}$ (a window of size $2\omega + 1$), as in Section 3.3.1 of Chapter 3. The coefficients $\{a_n\}$ are estimated using

$$a_n = \frac{\sum_{y' \in \Omega} Ch_n(y')\, g(i, j + y')}{\sum_{u \in \Omega} Ch_n^2(u)} \tag{4.4}$$

where $\{g(i, j + y')\}$ are the observed intensity values.

The first order derivatives of the intensity function at subpixel position $(i, j + y)$ can be calculated by

$$\frac{\partial g(i,j)}{\partial j}\Big|_{j=j+y} = \frac{d\hat{g}(i, j + y)}{dy} = \mathbf{A}^t \frac{d}{dy}\mathbf{CH}(y) \tag{4.5}$$

$$for \quad -0.5 \leq y < 0.5.$$

For simplicity of notation, we use $g'(i, j + y)$ to represent the first order partial derivatives of the subpixel intensity function.

For the purposes of performance analysis, rewrite the coefficient vector as

$$\mathbf{A} = \sum_{y' \in \Omega} \mathbf{A}(y')\, g(i, j + y') \tag{4.6}$$

where $\mathbf{A}(y')$ is a vector with the nth element

$$\frac{Ch_n(y')}{\sum_{u \in \Omega} Ch_n^2(u)}.$$

Replacing the vector $\mathbf{A}$ in (4.5) with (4.6) we have

$$g'(i, j + y) = \sum_{y' \in \Omega} F_y(y; y')\, g(i, j + y') \tag{4.7}$$

where $F_y(y; y')$ is given by

$$F_y(y; y') = \mathbf{A}^t(y')\frac{d}{dy}\mathbf{CH}(y). \tag{4.8}$$

The $F_y(y; y')$ can be considered as a filter for estimating the first order derivatives of the subpixel intensity function. Since only one camera is used, there does not exist bias amplitude in the observations. Suppose that the image is corrupted by zero mean white noise

$$\tilde{g}(i, j) = g(i, j) + n(i, j) \tag{4.9}$$

where $\tilde{g}(i, j)$ and $g(i, j)$ are the noisy observations and original intensity value, respectively. The variance is then given by

$$\mathbf{Var}(\sum_{y' \in \Omega} F_y(y; y')\tilde{g}(i, j + y'))$$

$$= \mathbf{E}\{(\sum_{y' \in \Omega} F_y(y; y')\, n(i, j + y'))^2\}$$

$$= \sigma_n^2 \sum_{y' \in \Omega} F_y^2(y; y') \tag{4.10}$$

where σ_n^2 is the variance of noise. Making use of the orthogonality of the polynomial set, we have

$$\sum_{y' \in \Omega} F_y^2(y; y') = \sum_{y' \in \Omega} (\mathbf{A}^t(y') \frac{d}{dy} \mathbf{CH}(y))^2$$

$$= (\frac{d}{dy} \mathbf{CH}^t(y)) \sum_{y' \in \Omega} \mathbf{A}(y') \, \mathbf{A}^t(y') \frac{d}{dy} \mathbf{CH}(y)$$

$$= (\frac{d}{dy} \mathbf{CH}^t(y)) \, \mathbf{W} \, \frac{d}{dy} \mathbf{CH}(y) \qquad (4.11)$$

where $\underline{\mathbf{W}}$ is a diagonal matrix with elements

$$w_{i,i} = \frac{1}{\sum_{u \in \Omega} Ch_i^2(u)}$$

$$\text{for} \quad i = 0, 1, ..., 4.$$

Thus, the variance becomes

$$\mathbf{Var}(\sum_{y' \in \Omega} F_y(y; y') \tilde{g}(i, j + y'))$$

$$= \sigma_n^2 \, (\frac{d}{dy} \mathbf{CH}^t(t)) \, \mathbf{W} \, \frac{d}{dy} \mathbf{CH}(y)$$

$$= \sigma_n^2 \sum_{n=1}^{4} \frac{(\frac{d}{dy} Ch_n(y))^2}{\sum_{v \in \Omega} Ch_n^2(v)}, \qquad (4.12)$$

which shows that the denominator of each term is a monotonically increasing function of the window size. Hence the variance of the output becomes smaller and smaller as the window size increases. Considering the effects of spatial quantization (as discussed in Chapter 3) and noise, a window with the size of three to seven pixels is recommended for natural images.

4.3.2 ESTIMATION OF CHAMFER DISTANCE VALUES

The Chamfer distance value is defined as the distance from the non-edge pixel to the nearest edge pixel [BTBW77]. To estimate Chamfer distance values, two steps are involved: converting the intensity image into a binary image consisting of the edge and non-edge pixels, and transforming the binary image to a Chamfer image.

Many conventional edge detectors can be used to find edge pixels from the intensity image. Since matching is restricted in the horizontal direction, only the horizontal Chamfer distance values are required. Computing horizontal Chamfer distance values needs some information about the vertical edges. For simplicity of implementation, we use the Prewitt edge detector [Pre70] with a window size of 3×3 to detect the vertical edges.

The Chamfer distance values are then iteratively determined using the following algorithm

$$f_{i,j}^{(l)} = min(f_{i,j-1}^{(l-1)} + 2, f_{i,j}^{(l-1)}, f_{i,j+1}^{(l-1)} + 2) \qquad (4.13)$$

where $f_{i,j}^{(\cdot)}$ is the distance value at the (i,j)th pixel and l denotes the iteration number. Initially, the distance values are set to zero for edge pixels and nonzero (say, 1000) for non-edge pixels. Edge pixels obviously get a value of zero. This algorithm is completely parallel and the iteration number is equal to the largest distance value occurring in the image.

4.4   Batch Approach

The conventional batch algorithm needs many measurements requiring a lot of computations. For example, if there are M image frames, then the matching procedure has to be implemented $(M-1)$ times to obtain $(M-1)$ disparity measurements for each pixel. Such a batch approach does not have any advantage over the recursive approach. Instead of doing matching $(M-1)$ times, the batch method presented in this section implements the matching algorithm only once by simultaneously using all the images.

Theoretically, disparity takes continuous values. For implementation purposes, we sample the disparity range using bins of size W. We use $N_r \times N_c \times (D+1)$ mutually interconnected binary neurons to represent the disparity measurements, where N_r and N_c are the image row and column sizes and $D \times W$ is the maximum disparity value. For the (i,j)th pixel, we use $(D+1)$ mutually exclusive neurons $\{v_{i,j,0}, v_{i,j,1}, ..., v_{i,j,D}\}$. When $v_{i,j,k}$ is 1, the disparity measured at pixel (i,j) is $k\,W$.

4.4.1   ESTIMATION OF PIXEL POSITIONS

Let (i,j) be the position of the (i,j)th pixel in the first frame. In successive frames, due to camera motion, the positions of all pixels are shifted to the right by vt. Under the epipolar constraint, the shift happens only in the horizontal direction. For example, the (i,j)th pixel moves from position (i,j) in the first frame to position $(i, j + \frac{fvt}{Z_{i,j}})$ in the second frame. Let $S_{i,j}(p)$ be the total shift of pixel (i,j) from the first frame to the pth frame. Thus

$$\begin{aligned} S_{i,j}(p) &= (p-1)\,\frac{fvt}{Z_{i,j}} \\ &= (p-1)\,d_{i,j} \qquad (4.14) \end{aligned}$$

where $d_{i,j}$ is the true disparity value for pixel (i,j). Note that the shift $S_{i,j}(p)$ is continuous due to the continuous variable $d_{i,j}$. A rounding operation has to be applied to $S_{i,j}(p)$ for locating the (i,j)th pixel in the subsampled image. After rounding, the position of the (i, j)th pixel in the pth frame is given by

$$(i, j + [\frac{S_{i,j}(p)}{W}]W) \qquad (4.15)$$

where [ ] is a rounding operator. It can be simply written as

$$(i, j + kW) \qquad (4.16)$$

where

$$k = [\frac{S_{i,j}(p)}{W}].$$

4.4.2 BATCH FORMULATION

Assuming that the camera is moving along the Y axis with constant velocity v and the images are taken exactly every t seconds apart, the error function for matching is defined as

$$
\begin{aligned}
E \;=\; & \frac{1}{M-1} \sum_{i=1}^{N_r} \sum_{j=1}^{N_c} \sum_{k=0}^{D} \sum_{p=1}^{M-1} [(g'_p(i, j + (p-1)kW) \\
& -g'_{p+1}(i, j + pkW))^2 + \kappa(f_p(i, j + (p-1)kW) \\
& -f_{p+1}(i, j + pkW))^2] \, v_{i,j,k} \\
& +\frac{\lambda}{2} \sum_{i=1}^{N_r} \sum_{j=1}^{N_c} \sum_{k=0}^{D} \sum_{s \in S} (v_{i,j,k} - v_{(i,j)+s,k})^2
\end{aligned}
\qquad (4.17)
$$

where $\{g'_p(\cdot)\}$ and $\{f_p(\cdot)\}$ denote the intensity derivatives and the Chamfer distance values at $(\cdot)$ of the pth frame, respectively. λ and κ are two constants. κ determines the relative importance of two kinds of measurement primitives. Comparison of (4.17) and (3.30) shows that the last terms in both equations are the same. Hence, the interconnection strength and bias input are given by

$$T_{i,j,k;l,m,n} = -48\lambda\delta_{i,l}\delta_{j,m}\delta_{k,n} + 2\lambda \sum_{s \in S} \delta_{(i,j),(l,m)+s}\delta_{k,n} \qquad (4.18)$$

and

$$
\begin{aligned}
I_{i,j,k} \;=\; & -\frac{1}{M-1} \sum_{p=1}^{M-1} [(g'_p(i, j + (p-1)kW) \\
& -g'_{p+1}(i, j + pkW))^2 + \kappa(f_p(i, j + (p-1)kW) \\
& -f_{p+1}(i, j + pkW))^2],
\end{aligned}
\qquad (4.19)
$$

respectively. It is interesting to note that using multiple images does not affect the interconnection strength, and the bias inputs contain all information about motion. The synchronous updating scheme described in Section

3.3 can be used for updating the states of neurons. The disparity values are determined by the stable states of neurons.

4.5 Recursive Approach

This approach basically consists of two steps: the bias input update and stereo matching. Whenever a new frame of image becomes available, the bias inputs of the network are updated by the RLS algorithm and then new disparity measurements are obtained using these new bias inputs.

Suppose that images are corrupted by additive white noise and the measurement model is given by

$$
\begin{aligned}
\tilde{I}_{i,j,k}(p) &= h(p, g_p'(i, j + (p-1)kW), f_p(i, j + (p-1)kW), \\
&\quad g_{p+1}'(i, j + pkW), f_{p+1}(i, j + pkW), n_{i,j,k}(p)) \\
&= -(\tilde{g}_p'(i, j + (p-1)kW) - \tilde{g}_{p+1}'(i, j + pkW))^2 \\
&\quad -\kappa(\tilde{f}_p(i, j + (p-1)kW) - \tilde{f}_{p+1}(i, j + pkW))^2
\end{aligned}
$$

$$
\text{for} \quad p = 1, 2, ..., M-1
$$

where h is a measurement function and $n_{i,j,k}(p)$ is noise. For p given measurements $\{\tilde{I}_{i,j,k}(p), \tilde{I}_{i,j,k}(p-1), ..., \tilde{I}_{i,j,k}(1)\}$, we want to find a function

$$
\hat{I}_{i,j,k}(p) = \hat{I}_{i,j,k}(p, \tilde{I}_{i,j,k}(p), \tilde{I}_{i,j,k}(p-1), ..., \tilde{I}_{i,j,k}(1)), \tag{4.20}
$$

which gives the estimate of the bias input $I_{i,j,k}$. If the measurement function is linear and the measurement noise is white, then a Kalman filter is commonly used for finding an optimal estimate. In (4.20), as the measurement function is nonlinear and the measurement noise is no longer white but is dependent on measurements, the linear Kalman filter does not yield a good estimate. In contrast to the Kalman filter, the RLS algorithm does not make any assumption about measurement function and noise. Hence, the RLS algorithm can be used to update the bias inputs. When the pth frame becomes available, the bias input is updated by

$$
I_{i,j,k}(p) = I_{i,j,k}(p-1) + \frac{1}{p}\left(\tilde{I}_{i,j,k}(p) - I_{i,j,k}(p-1)\right). \tag{4.21}
$$

This RLS algorithm is equivalent to the batch least squares algorithm with the initial condition

$$
I_{i,j,k}(0) = 0.
$$

The matching algorithm is the same as that used for static stereo matching. The interconnection strengths are the same as in (3.33):

$$
T_{i,j,k;l,m,n} = -48\lambda\delta_{i,l}\delta_{j,m}\delta_{k,n} + 2\lambda\sum_{s\in S}\delta_{(i,j),(l,m)+s}\delta_{k,n}, \tag{4.22}
$$

and the bias input is given by (4.21). Since the bias inputs are recursively updated and contain all the information about the previous images, we do not need to implement the matching algorithm for every recursion if the intermediate results are not required. This method greatly reduces the computational load and therefore is extremely fast. Formally, the algorithm is as follows:

1. Update the bias inputs using the RLS algorithm.

2. Initialize the neuron states.

3. If there is a new frame to be processed, go back to step 1; otherwise go to step 4.

4. Update the neuron states using the matching algorithm.

This algorithm has several advantages over the correlation algorithm of [MSK88]:

1. This algorithm recursively updates the bias inputs instead of the disparity values. The matching algorithm is implemented only once.

2. Instead of using an extra smoothing procedure, this algorithm incorporates the smoothness constraint into the matching procedure.

3. This algorithm uses the derivatives of the intensity function, which are more reliable than the intensity values, as measurement primitives. Hence it is suitable for natural images.

4.6   Matching Error

When multiple frames are used for matching, the spatial quantization error usually causes a large matching error. In this section, we derive a mean value for the matching error which can be used to detect the occluding pixels. It is assumed that the true disparity value at pixel (i, j) in a smooth region can be expressed as

$$d_{i,j} = kw + \delta_{i,j}$$

where $\delta_{i,j}$ is uniformly distributed in $[-\frac{w}{2}, \frac{w}{2})$, and the first order derivative of the intensity function at point $(i, j + (p-1)kw)$ of the pth image can be expanded as a Taylor series about the point $(i, j + (p-1)(kw + \delta_{i,j}))$ as

$$
\begin{aligned}
g'(i, j + (p-1)kw) \;=\; & g'(i, j + (p-1)(kw + \delta_{i,j})) \qquad\qquad (4.23) \\
& - (p-1)\delta_{i,j}g''(i, j + (p-1)(kw + \delta_{i,j})) + o(\delta_{i,j}^2)
\end{aligned}
$$

$$\text{for} \quad p = 2, 3, ..., m$$

where $g''(\cdot)$ denotes the second order derivative. The best estimate of the disparity value is given by

$$\hat{d}_{i,j} = kw.$$

Therefore, the matching error is

$$error = \frac{1}{m-1} \sum_{p=1}^{m-1} [g_p'(i, j + (p-1)kw) - g_{p+1}'(i, j + pkw)]^2$$

$$\simeq \frac{1}{m-1} \sum_{p=1}^{m-1} [g_p'(i, j + (p-1)(kw + \delta_{i,j}))$$

$$-g_{p+1}'(i, j + p(kw + \delta_{i,j}))$$

$$-(p-1)\delta_{i,j} g_p''(i, j + (p-1)(kw + \delta_{i,j}))$$

$$+p\,\delta_{i,j} g_{p+1}''(i, j + p(kw + \delta_{i,j}))]^2. \tag{4.24}$$

When images are corrupted by additive white noise

$$\tilde{g}_p(i, j) = g_p(i, j) + n_p(i, j)$$

where $\tilde{g}_p(i, j)$ is the observation, the first order derivative of the intensity function is given by

$$\tilde{g}_p'(i, j + y) = \sum_{y' \in \omega} \mathbf{a}^t(y') \frac{d}{dy} \underline{\mathbf{ch}}(y)\, \tilde{g}_p(i, j + y') \tag{4.25}$$

$$= g_p'(i, j + y) + \sum_{y' \in \omega} \mathbf{a}^t(y') \frac{d}{dy} \underline{\mathbf{ch}}(y)\, n_p(i, j + y')$$

$$\text{for} \quad -0.5 \le y < 0.5$$

and the second order derivative is given by

$$\tilde{g}_p''(i, j + y) = \sum_{y' \in \omega} \mathbf{a}^t(y') \frac{d^2}{dy^2} \underline{\mathbf{ch}}(y)\, \tilde{g}_p(i, j + y') \tag{4.26}$$

$$= g_p''(i, j + y) + \sum_{y' \in \omega} \mathbf{a}^t(y') \frac{d^2}{dy^2} \underline{\mathbf{ch}}(y)\, n_p(i, j + y')$$

$$\text{for} \quad -0.5 \le y < 0.5.$$

Replacing $g_p'(\cdot)$ and $g_p''(\cdot)$ in (4.24) by (4.25) and (4.26) and noting that

$$g_1'(i, j) = g_p'(i, j + (p-1)(kw + \delta_{i,j}))$$

and

$$g_1''(i, j) = g_p''(i, j + (p-1)(kw + \delta_{i,j}))$$

$$\text{for} \quad p = 2, 3, ..., m,$$

the matching error becomes

$$
\begin{aligned}
error \quad = \quad & \frac{1}{m-1} \sum_{p=1}^{m-1} [\sum_{y' \in \omega} \mathbf{a}^t(y') \frac{d}{dy} \underline{\mathbf{ch}}(y)|_{y=\bar{y}_{p-1}} n_p(i, j + y_{p-1} + y') \\
& - \sum_{y' \in \omega} \mathbf{a}^t(y') \frac{d}{dy} \underline{\mathbf{ch}}(y)|_{y=\bar{y}_p} n_{p+1}(i, j + y_p + y') \\
& - (p-1)\delta_{i,j} \sum_{y' \in \omega} \mathbf{a}^t(y') \frac{d^2}{dy^2} \underline{\mathbf{ch}}(y)|_{y=\bar{y}_{p-1}} n_p(i, j + y_{p-1} + y') \\
& + p\, \delta_{i,j} \sum_{y' \in \omega} \mathbf{a}^t(y') \frac{d^2}{dy^2} \underline{\mathbf{ch}}(y)|_{y=\bar{y}_p} n_{p+1}(i, j + y_p + y') \\
& + \delta_{i,j}\, g_1''(i, j)]^2
\end{aligned}
$$

$$(4.27)$$

where

$$y_p = [pkw]$$

$$\bar{y}_p = pkw - y_p$$

and $[\ ]$ is a rounding operator. Assuming that noise $\{n_p(i, j)|p = 1, 2, ..., m\}$ and $\delta_{i,j}$ are mutually independent and noting the orthogonality of the polynomial set, the mean error can be simplified as

$$
\begin{aligned}
\mathbf{e}\{error\} \quad = \quad & \frac{1}{m-1} \sum_{p=1}^{m-1} \{\sum_{n=1}^{4} \frac{1}{\sum_{v \in \omega} ch_n^2(v)} \; [\sigma_{n_p}^2 (\frac{d}{dy} ch_n(y))^2|_{y=\bar{y}_{p-1}} \\
& + \sigma_{n_{p+1}}^2 (\frac{d}{dy} ch_n(y))^2|_{y=\bar{y}_p} \\
& + \frac{w^2 p^2 \sigma_{n_{p+1}}^2}{12} (\frac{d^2}{dy^2} ch_n(y))^2|_{y=\bar{y}_p} \\
& \frac{w^2 (p-1)^2 \sigma_{n_p}^2}{12} (\frac{d^2}{dy^2} ch_n(y))^2|_{y=\bar{y}_{p-1}}] \\
& + \frac{w^2}{12} (g_1''(i, j))^2\}.
\end{aligned}
$$

$$(4.28)$$

Since only one camera is used, it is reasonable to assume that

$$\sigma_{n_1}^2 = \sigma_{n_p}^2 \quad \text{for} \ p = 2, ..., m.$$

The mean error at (i, j) can be computed as

$$\mathbf{e}\{error(i, j)_k\} = c_1 \sigma_{n_1}^2 + c_2 (g_1''(i, j))^2$$

$$(4.29)$$

where c_1 and c_2 are determined by

$$c_1 = \frac{1}{m-1} \sum_{p=1}^{m-1} \{ \sum_{n=1}^{4} \frac{1}{\sum_{v \in \omega} ch_n^2(v)} [(\frac{d}{dy} ch_n(y))^2 |_{y=\bar{y}_{p-1}}$$

$$+ (\frac{d}{dy} ch_n(y))^2 |_{y=\bar{y}_p}] + \frac{w^2}{12} \sum_{n=2}^{4} [p^2 (\frac{d^2}{dy^2} ch_n(y))^2 |_{y=\bar{y}_p}$$

$$+ (p-1)^2 (\frac{d^2}{dy^2} Ch_n(y))^2 |_{y=\bar{y}_{p-1}}]\} \qquad (4.30)$$

and

$$c_2 = \frac{W^2}{12}, \qquad (4.31)$$

respectively. If the variance of noise and the second order derivatives of the intensity function are known or estimated from images, then the mean error corresponding to the disparity value k at each point for given ω (window size), M (frame number) and W (width of subsample interval) can be calculated.

4.7   Detection of Occluding Pixels

Detection of occluding pixels is an important issue in motion stereo. As shown in Figure 4.2(a), when a camera moves from right to left, points 2, 3, 4, and 5 project into the first image plane at 2', 3', 4', and 5'. However, points 2 and 3 on the object surface will not project into the second image plane because they are occluded by the front surface on which point 1 lies. Similarly, points 2, 3, 4, and 5 will not appear in the image plane 3. At the location of pixels 2', 3', 4', and 5', the matching error is usually large which means no conjugate points can be found in the successive image frames. Hence, the disparity values are undetermined. Such pixels are called occluding pixels. When a smoothness constraint is used, although the matching algorithm always assign some values to the occluding pixels, the discontinuities of the disparity field may be shifted. As the number of frames increases, the number of occluding pixels also increases dramatically. For instance, if only two object points are occluded for the second image as shown in Figure 4.2(a), then about ten points are occluded for the sixth image which gives a ten pixel wide occluding region in the first image plane. On the other hand, if only the first two frames are used for matching, then pixels 4' and 5' are not occluding pixels and therefore the disparity values at such location are determinable. As the third frame does not provide any information about pixels 4' and 5', there is no need to update the bias inputs at the location of these pixels.

However, in some cases the number of occluding pixels does not increase as the number of frames increases. One typical example is illustrated in

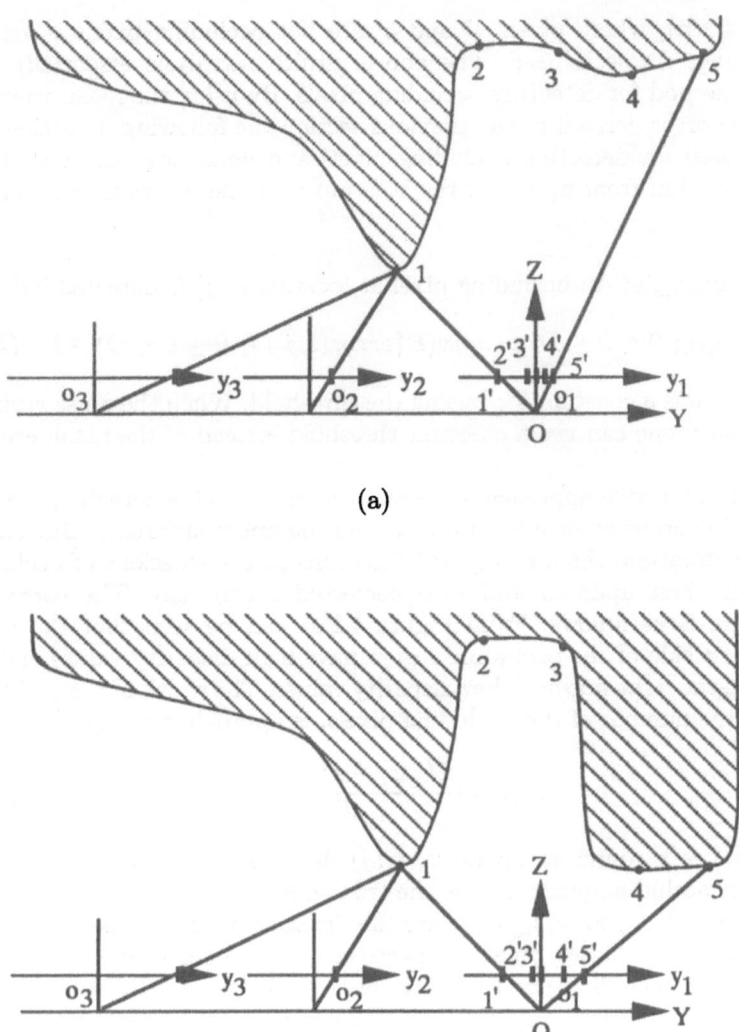

FIGURE 4.2. Occluding pixels.(a) Pixels 2', 3', 4', and 5' are occluded. (b) Pixels 4' and 5' are not occluded.

Figure 4.2(b), where pixels 4' and 5' are not occluding pixels when the third image frame is used. The above intuitive analysis essentially suggests a method for detecting occluding pixels. By using the mean values of matching error derived in the previous section the following detection rule can be used for detecting occluding pixels and hence we can prevent the RLS algorithm from updating the bias input at the location of occluding pixels.

Detection rule: An occluding pixel at location (i, j) is detected if

$$min(I_{i,j,k};\ 0 \le k \le D) > max(\mathbf{E}\{error(i,j)_k\};\ 0 \le k \le D) + b \quad (4.32)$$

where $b \ge 0$ is a constant for raising the threshold. When the noise variance is unknown, one can use a constant threshold instead of the mean error.

For the recursive approach, once an occluding pixel is detected, the bias inputs of neurons at such locations will not be updated further. But during the first iteration, the bias inputs of neurons at the locations of occluding pixels are first updated and then corrected accordingly. The correction procedure is as follows. From Figure 4.2 it can be seen that the pixels on the left side of the occluding region have high disparity values and the pixels on the right side have low disparity values. The width of the occluding region, the number of the occluding pixels, is approximately given by

$$\hat{\Delta}_{i,j} = [\frac{D_{i,j-1} - D_{i,j+\Delta_{i,j}}}{W}] \quad (4.33)$$

where $[\ ]$ is a rounding operator, (i, j) denotes the location of the farthest left occluding pixel, $\Delta_{i,j}$ is the true width, $\hat{\Delta}_{i,j}$ is an estimate of the width, and $D_{i,j-1}$ and $D_{i,j+\Delta_{i,j}}$ are the disparity values of the nearest left and right nonoccluding pixels, respectively. Since a smoothness constraint is used, the occluding pixel usually takes either the high disparity value $D_{i,j-1}$ or low disparity value $D_{i,j+\Delta_{i,j}}$. The discontinuities of the disparity field can be detected by checking the disparity values in the y_1 direction for a transition from the high value to the low value. Starting with the discontinuity pixel, we check all the neighboring pixels to the left and right. The search procedure will not be stopped until a $\hat{\Delta}_{i,j}$ wide or less occluding region including the discontinuity pixel is found. Then, for all occluding pixels, the bias input of the $\Delta_{i,j}$th neuron is corrected by

$$I_{i,l,D_{i,j+\Delta_{i,j}}}(1) = min(I_{i,l,k}(1);\ k = 0, 1, ..., D). \quad (4.34)$$

For the batch approach, the bias inputs at occluding pixels are estimated using only the first two image frames.

4.8   Experimental Results

We have tested both the batch and recursive algorithms on several sequences of natural images taken by sliding a camera from right to left. Figure 4.3 shows three frames of a tree sequence of 120 frames. The size of the images is 256×233. We arbitrarily chose five successive frames, although there is no limit to the number of frames. No alignment in the vertical direction was made and the maximum disparity, approximately two pixels, was measured by hand. The same parameters were chosen for both algorithms. W was set at 0.2 and hence $D = 10$. The parameters λ and κ were set at 20 and 5, respectively. For detecting an occluding pixel, the threshold was set at 150 because the noise variance is unknown. Figure 4.4 shows the batch result and its smooth version after 45 iterations. The disparity map is represented by an intensity image. Since at occluding regions we still use the measurement primitives extracted from the first two image frames for matching, the algorithms might generate some isolated points or regions (at most $[WD]$ pixels wide) due to the incorrect information caused by the occlusions. To remove such points and regions, a median filter is used in our experiments. The recursive result and its smooth version are shown in Figure 4.5. The 20 iterations are required for the recursive solution.

4.9   Discussion

We have presented two methods, the batch and recursive, based on the neural network for lateral motion stereo. The first order derivatives of intensity function are used for measurement primitives. In the recursive approach, the bias inputs of the neurons are recursively updated. If the intermediate results are not required, the matching procedure is implemented only once. Unlike the existing recursive approach, the disparity field obtained by this approach is smooth and dense. In addition, no batch results are needed for setting the initial states of the neurons. Both batch and recursive methods gave very good results in comparison to Barnard's approach [Bar86]. Experimental results show that the recursive approach needs fewer iterations than the batch approach. This is because the recursive approach uses a better bias input updating scheme (especially for the occluding pixels). The good estimate of the bias inputs makes the network converge quickly, although the updating step for bias inputs takes more computations. In view of parallelism and fast convergence, the recursive approach is useful for real time implementation, such as in a robot vision system. In our experiment, the threshold used was 150, which seems a little bit conservative. However, the maximum disparity is only approximately two pixels, so the width of the occluding region is less than two pixels for two frames and there are only a few occluding pixels along the right boundaries of the trees. Hence the occluding pixels do not cause a serious problem in this experiment.

This is also why the iteration number does not decrease a lot. We believe that if the maximum disparity is large and a long sequence of images is used, then the improvement on the occluding pixel detection will greatly reduce the number of iterations. Also note that since our algorithms locate occluding pixels accurately, using more image frames does not increase the number of the undeterminable pixels. Instead it gives information on the occluding pixels which can be used to locate the discontinuities of the depth map. Although our algorithms require constant velocity camera motion and equally spaced images, these requirements are unnecessary. For instance, if the velocity of the camera is known, then Equation (4.14) can written as

$$S_{i,j}(p) = \frac{f}{z_{i,j}} \sum_{i=0}^{p-1} (t_i - t_{i-1})\, v_i \qquad (4.35)$$

where v_i is the camera velocity at time t_i. Thus, the total shift of the (i,j)th pixel from the first frame to the pth frame can be determined by using (4.35) rather than (4.14). The equally spacing requirement can be removed.

(a)   (b)

(c)

FIGURE 4.3. The Trees motion sequence. (a) The first frame. (b) The second frame. (c) The fifth frame.

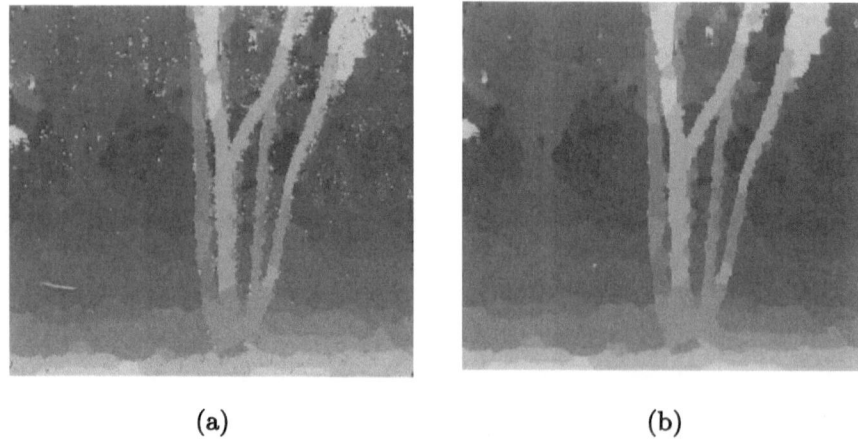

(a)   (b)

FIGURE 4.4. Batch approach. (a) Disparity map. (b) Smoothed version of the disparity map. (Reprinted with permission from [ZC89] Figure 3, ©1989 IEEE.)

(a)   (b)

FIGURE 4.5. Recursive approach. (a) Disparity map. (b) Smoothed version of the disparity map. (Reprinted with permission from [ZC89] Figure 3, ©1989 IEEE.)

5

Motion Stereo–Longitudinal Motion

5.1 Introduction

Longitudinal motion stereo infers depth information from a forward or backward motion, and consequently is particularly useful in autonomous robot navigation applications. Most existing algorithms have some problems associated with the location of the FOE, and with the camera and surface orientations. These problems limit their applicability to real scenes. It is highly desirable to develop an efficient and practical algorithm to infer depth information from longitudinal motion. In this chapter, we present a neural network-based algorithm for longitudinal motion stereo. The algorithm allows the camera to move along its optical axis forward or backward, and requires no information on the FOE. It produces multiple dense disparity fields and recovers the depth map very efficiently.

In the 1980s, Williams [Wil80] presented an approach for deriving depth from a forward moving camera. By moving a camera forward, disparity is estimated by using simple triangulation. For simplicity, all object surfaces are assumed to be flat and oriented in either the horizontal direction, parallel to the image plane, or the vertical direction, parallel to the ground plane. Therefore, only the distances from the camera to the horizontal surface and the height of the vertical surface need to be found. To achieve subpixel accuracy, an image is interpolated according to the predicted disparity values obtained by a search process and occlusion effects. Based on the error between the real and interpolated images, the correct orientation of each surface is detected, and hence a segmented image consisting of refined synthetic surfaces is obtained. For implementation purposes, an iterative segmentation procedure is employed, and the systematic changes of distance and height embodied in the synthetic segmented image at each iteration are used for finding the correct distance and height. Experimental results demonstrate the usefulness of this approach for simple natural image sequences. Since only planar surfaces oriented in the horizontal direction or the vertical direction are assumed to exist in natural images, areas corresponding to either non-planar surfaces or planar surfaces with other orientations are not correctly interpolated, and therefore the estimated distances and heights for these areas are not reliable. Furthermore, this approach requires information about the FOE and the final result depends very much on the quality of initial segmentation.

Bolles and Baker's approach [BB85] computes depth from the object path through the epipolar-plane images (EPIs). Their approach slices the spatial-temporal data along a temporal dimension, locates features in these slices, and computes dimensional locations. The structured temporal images, that are epipolar-plane images, are built by assuming that the camera moves in a straight line laterally or forward. But in forward motion, the optical axis of the camera is not parallel to the direction of motion exactly. Instead, the camera is at a fixed angle relative to the direction of motion. The angle value is assumed to be known.

In order to find an efficient and natural way to compute depth from motion pictures, we assume that the camera moves forward or backward along its optical axis and that the origin of the image plane coordinate system is at the center of the image. Centering the image plane coordinate system makes the FOE coincident with the origin of the system. There is no need to locate the FOE anymore. Since our algorithm makes no assumption for the surfaces except smoothness, it is able to handle any planar or non-planar surfaces. Since the major difference between the forward and backward motions is a matter of image order in the sequence, the forward motion stereo algorithm can be used for backward motion by simply reversing the image order. Motion stereo algorithms are commonly based on the assumption that camera velocity is known. Even though this assumption is not always stated, it is universal.

One common problem with motion stereo is the measurement primitives. Using the image intensity and its derivatives as measurement primitives suffers from noise distortion since real images are usually corrupted by some noise. Meanwhile, using image features, such as edges and corners, as the measurement primitives requires a surface interpolation step to obtain a dense depth map since edges and corners are usually sparse, and depth values only at these feature points can be recovered. To avoid a surface interpolation step and make the algorithm more robust, our algorithm uses the Gabor features as the measurement primitives. The Gabor features are obtained by applying a Gabor correlation operator to each image point. The Gabor correlation operator consists of a set of multiresolution two-dimensional Gabor functions. The reason for using the Gabor correlation operator is based on the evidence that two-dimensional Gabor functions fit the two-dimensional receptive profiles of the mammalian cortical simple cells, and their parameters capture the neurophysiological properties of localization in visual space, spatial dimensions, preferred orientation and spatial frequency [Dau85]. Gabor functions can be considered as the receptive field of the neurons. Based on the multiresolution Gabor features, our algorithm employs a discrete parallel neural network to compute the disparity field. To take full advantage of the computationally cooperative property of the network, some smoothness constraints are embedded into the network. Finally, the algorithm recovers a dense depth map from the computed disparity fields by using a simple algebraic method. Our algorithm has been successfully applied to real motion images.

5.2 Depth from Forward Motion

5.2.1 GENERAL CASE: IMAGES ARE NONEQUALLY SPACED

Figure 5.1 shows a camera geometry for forward motion. A camera is moving forward at velocity v relative to a stationary scene. The optical axis of the camera is assumed to be parallel to the moving direction. Image planes are assumed to be perpendicular to the moving direction. Let $OXYZ$ be the space coordinate system with the Z axis oriented along the optical axis of the camera, $o_i x_i y_i$ be the ith image plane coordinate system, f be the focal length of the lens which takes a positive value in the space system, and $P = (X_o, Y_o, Z_o)$ be a point in space. Under perspective projection, P projects onto a point P_i on the ith image plane at (x_{Pi}, y_{Pi}) which can be determined by using the similarity of the triangles. As shown in Figure 5.2, since triangle $c_i V Z_o$ is similar to triangle $c_i x_{Pi} o_i$, and triangle $c_i H Z_o$ is similar to triangle $c_i y_{Pi} o_i$, we have

$$\frac{x_{Pi}}{f} = \frac{X_o}{Z_o - (t_i - t_1)\, v} \tag{5.1}$$

and

$$\frac{y_{Pi}}{f} = \frac{Y_o}{Z_o - (t_i - t_1)\, v}. \tag{5.2}$$

The position of point P_i is then given by

$$(x_{Pi}, y_{Pi}) = \left(\frac{f\, X_o}{Z_o - (t_i - t_1)\, v}, \frac{f\, Y_o}{Z_o - (t_i - t_1)\, v} \right). \tag{5.3}$$

Horizontal and vertical disparities can be computed from $(i-1)$th and ith frames by

$$\begin{aligned} D_{x_{i,i-1}} &= x_{Pi} - x_{P(i-1)} \\ &= \frac{(t_i - t_{i-1})\, v\, f\, X_o}{(Z_o - (t_i - t_1)\, v)(Z_o - (t_{i-1} - t_1)\, v)} \end{aligned} \tag{5.4}$$

and

$$\begin{aligned} D_{y_{i,i-1}} &= y_{Pi} - y_{P(i-1)} \\ &= \frac{(t_i - t_{i-1})\, v\, f\, Y_o}{(Z_o - (t_i - t_1)\, v)(Z_o - (t_{i-1} - t_1)\, v)}, \end{aligned} \tag{5.5}$$

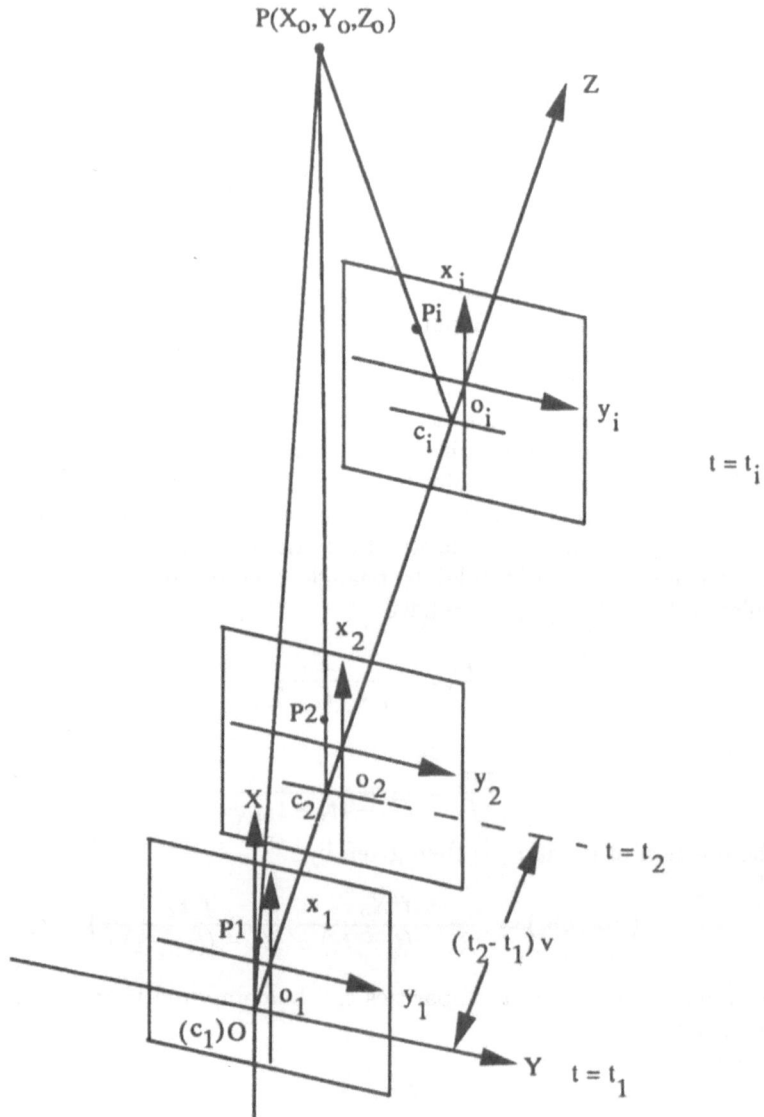

FIGURE 5.1. Camera geometry for forward motion.

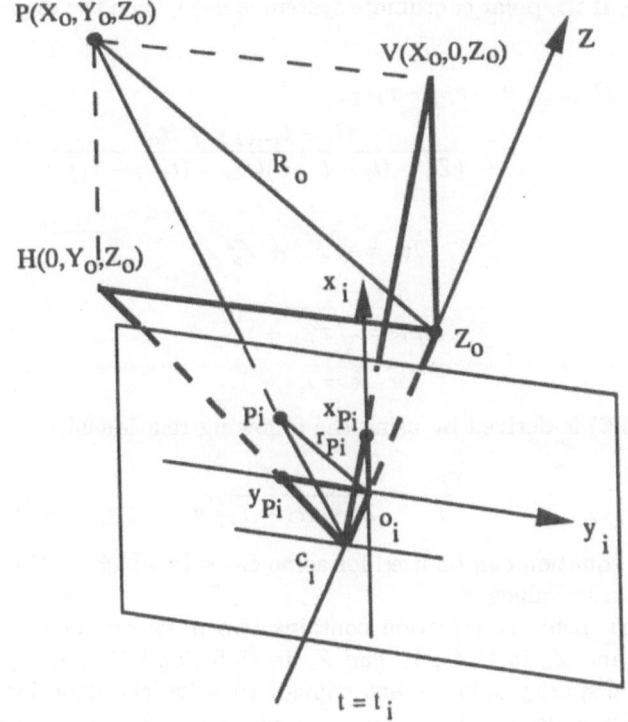

FIGURE 5.2. Similar triangles. Triangle $C_i X_o Z_o$ is similar to triangle $c_i x_{Pi} o_i$, and triangle $C_i Y_o Z_o$ is similar to triangle $c_i y_{Pi} o_i$.

respectively. If the polar coordinate system is used, then the disparity value is given by

$$
\begin{aligned}
D_{i,i-1} &= r_{Pi} - r_{P(i-1)} \\
&= \frac{(t_i - t_{i-1})\, v\, f\, R_o}{(Z_o - (t_i - t_1)\, v)(Z_o - (t_{i-1} - t_1)\, v)}
\end{aligned}
\tag{5.6}
$$

where

$$
R_o = \sqrt{X_o^2 + Y_o^2}
\tag{5.7}
$$

and

$$
r_{Pk} = \sqrt{x_{Pk}^2 + y_{Pk}^2}
\tag{5.8}
$$

$$
\text{for} \quad k = i, i - 1.
$$

Equation (5.6) is derived by using the following relationship:

$$
\frac{r_{Pk}}{f} = \frac{R_o\, f}{Z_o - (t_k - t_1)\, v}.
\tag{5.9}
$$

This useful equation can be used for some cases in which (5.4) or (5.5) fail to find disparity values.

Since each disparity equation contains two unknown variables (for instance, X_o and Z_o in (5.4), Y_o and Z_o in (5.5), and R_o and Z_o in (5.6)), one known disparity value is not enough to solve the equation. At least two disparity values or three frames are needed to determine depth Z_o. It is interesting to note that lateral motion needs only two frames to recover depth value [ZC89]. This is because the Z direction distance from the image plane to the surface is a constant. As shown in (4.2), subtracting y_{P-1} from y_P eliminates Y_o. In forward motion, the Z direction distance is a function of time t. From (5.5) it can be seen that subtracting $Y_{P(i-1)}$ from Y_{Pi} does not eliminate Y_o.

When given $D_{x_{i+1,1}}$ and $D_{x_{i,i-1}}$, depth Z_o can be computed by the following equation:

$$
Z_o = \frac{[D_{x_{i,i-1}}(t_{i+1} - t_i)(t_{i-1} - t_1) - D_{x_{i+1,i}}(t_i - t_{i-1})(t_{i+1} - t_1)]v}{D_{x_{i,i-1}}(t_{i+1} - t_i) - D_{x_{i+1,i}}(t_i - t_{i-1})}.
\tag{5.10}
$$

The above equation is derived from (5.4) using simple algebraic manipulations. Depth Z_o can also be recovered from $D_{y_{i+1,1}}$ and $D_{y_{i,i-1}}$ by

$$
Z_o = \frac{[D_{y_{i,i-1}}(t_{i+1} - t_i)(t_{i-1} - t_1) - D_{y_{i+1,i}}(t_i - t_{i-1})(t_{i+1} - t_1)]v}{D_{y_{i,i-1}}(t_{i+1} - t_i) - D_{y_{i+1,i}}(t_i - t_{i-1})},
\tag{5.11}
$$

or from $D_{i+1,1}$ and $D_{i,i-1}$ by

$$
Z_o = \frac{[D_{i,i-1}(t_{i+1} - t_i)(t_{i-1} - t_1) - D_{i+1,i}(t_i - t_{i-1})(t_{i+1} - t_1)]v}{D_{i,i-1}(t_{i+1} - t_i) - D_{i+1,i}(t_i - t_{i-1})}.
\tag{5.12}
$$

It should be noted that a point on the $Y - Z$ plane has the zero disparity value in the horizontal direction since $X_o = 0$ leads to $D_{x_i,i-1} = 0$. As a consequence, (5.10) is not valid for this point. To recover the depth value for this point, we have to first use (5.5) to estimate the disparity values and then use (5.11) to compute Z_o. Similarly, (5.11) is not valid for the point on the $X - Z$ plane. By contrast, (5.12) is valid for the point on the $X - Z$ plane or $Y - Z$ plane, but not for a point on both $X - Z$ and $Y - Z$ planes, that is on the Z axis. When a point is on the Z axis, it projects onto the FOE (the origin), which has zero disparity values in all directions. Since none of (5.4), (5.5), and (5.6) can be used for this point, we use its neighboring points to determine the depth value by using an interpolation technique under the assumption that its neighboring points and itself are from a smooth surface.

It should also be noted that using $D_{x_i,i-1}$ and $D_{x_{i+1},i}$ to determine depth Z_o cause an accumulated error. Because images are usually spatially quantized. The computation of $D_{x_i,i-1}$ requires that a corresponding point (x_{Pi}, y_{Pi}) in the ith frame be found based on a point $(x_{P(i-1)}, y_{P(i-1)})$ in the $(i - 1)$th frame. Similarly, the computation of $D_{x_{i+1},i}$ requires the location of a second corresponding point $(x_{P(i+1)}, y_{P(i+1)})$ in the $(i + 1)$th frame based on the first corresponding point (x_{Pi}, y_{Pi}) in the ith frame. If the first corresponding point (x_{Pi}, y_{Pi}) is in some place between two adjacent sample points in the ith frame, then not only the first corresponding point (x_{Pi}, y_{Pi}) but also the second corresponding point $(x_{P(i+1)}, y_{P(i+1)})$ cannot be located accurately due to the accumulated error, even though a subpixel interpolation step has been involved. To avoid the accumulated error, we use $D_{x_{i+1},i-1}$ instead of $D_{x_{i+1},i}$ to compute depth Z_o. Replacing $D_{x_{i+1},i-1}$ with $D_{x_{i+1},i}$ in (5.10) gives

$$Z_o = \frac{[D_{x_i,i-1}(t_{i+1} - t_{i-1})(t_i - t_1) - D_{x_{i+1},i-1}(t_i - t_{i-1})(t_{i+1} - t_1)]v}{D_{x_i,i-1}(t_{i+1} - t_{i-1}) - D_{x_{i+1},i-1}(t_i - t_{i-1})}$$

(5.13)

where $D_{x_{i+1},i-1}$ is given by

$$\begin{aligned} D_{x_{i+1},i-1} &= x_{P(i+1)} - x_{P(i-1)} \\ &= \frac{(t_{i+1} - t_{i-1})\, v\, f\, X_o}{(Z_o - (t_{i+1} - t_1)\, v)(Z_o - (t_{i-1} - t_1)\, v)}. \end{aligned}$$

(5.14)

To estimate $D_{x_{i+1},i-1}$, $(x_{P(i-1)}, y_{P(i-1)})$ and $(x_{P(i+1)}, y_{P(i+1)})$ are needed. The second corresponding point is then located based on $(x_{P(i-1)}, y_{P(i-1)})$, rather than upon the first corresponding point. Similarly, (5.11) and (5.12) can be modified as

$$Z_o = \frac{[D_{y_i,i-1}(t_{i+1} - t_{i-1})(t_i - t_1) - D_{y_{i+1},i-1}(t_i - t_{i-1})(t_{i+1} - t_1)]v}{D_{y_i,i-1}(t_{i+1} - t_{i-1}) - D_{y_{i+1},i-1}(t_i - t_{i-1})}$$

(5.15)

and

$$Z_o = \frac{[D_{i,i-1}(t_{i+1} - t_{i-1})(t_i - t_1) - D_{i+1,i-1}(t_i - t_{i-1})(t_{i+1} - t_1)]v}{D_{i,i-1}(t_{i+1} - t_{i-1}) - D_{i+1,i-1}(t_i - t_{i-1})}.$$

(5.16)

When images are out of alignment, the following equation can be used to align them:

$$\frac{D_{x_{i,i-1}}}{D_{y_{i,i-1}}} = \frac{x_{Pi}}{y_{Pi}}.$$

(5.17)

The relationship is obtained by dividing $D_{x_{i,i-1}}$ by $D_{y_{i,i-1}}$ and replacing X_o with (5.1) and Y_o with (5.2). In (5.17), we assume that either $D_{x_{i,i-1}}$ or $D_{y_{i,i-1}}$ is not a zero.

5.2.2   SPECIAL CASE: IMAGES ARE EQUALLY SPACED

When images are equally spaced, all the equations can be further simplified. Assuming images are taken exactly every t seconds apart, the time variable t can be set as

$$t_i = (i - 1)t, \quad \text{for} \quad i = 1, 2, ...$$

As a consequence, using (5.4) $D_{x_{i,i-1}}$ can be written as

$$
\begin{aligned}
D_{x_{i,i-1}} &= x_{Pi} - x_{P(i-1)} \\
&= \frac{t \, v \, f \, X_o}{(Z_o - (i-1)t \, v)(Z_o - (i-2)t \, v)},
\end{aligned}
$$

(5.18)

and $D_{x_{i+1,i}}$ can be written as

$$
\begin{aligned}
D_{x_{i+1,i}} &= x_{P(i+1)} - x_{Pi} \\
&= \frac{t \, v \, f \, X_o}{(Z_o - i \, t \, v)(Z_o - (i-1)t \, v)}.
\end{aligned}
$$

(5.19)

Since $Z_o - (i-2)t \, v > Z_o - i \, t \, v$,

$$D_{x_{i+1,i}} > D_{x_{i,i-1}},$$

(5.20)

and the disparity becomes larger and larger as the camera moves forward. This represents a condition on the estimated disparity values. In the lateral motion case, the disparity values are constant provided the images are equally spaced [ZC89]. Although (5.20) is useful for reducing the search range for stereo matching, it is a major obstacle for deriving a multiple (more than three) frame algorithm for longitudinal motion stereo.

Combining (5.18) and (5.19), we obtain

$$Z_o = t \, v \, (i + \frac{2 \, D_{x_{i,i-1}}}{D_{x_{i+1,i}} - D_{x_{i,i-1}}}).$$

(5.21)

This is a simplified version of (5.10). In a similar manner, using $D_{y_{i+1,i}}$ and $D_{y_{i,i-1}}$,

$$Z_o = t\,v\,(i + \frac{2\,D_{y_{i,i-1}}}{D_{y_{i+1,i}} - D_{y_{i,i-1}}}) \tag{5.22}$$

and using $D_{i+1,i}$ and $D_{i,i-1}$

$$Z_o = t\,v\,(i + \frac{2\,D_{i,i-1}}{D_{i+1,i} - D_{i,i-1}}). \tag{5.23}$$

To reduce the effect of the accumulated error, we use (5.13) to compute the depth values. With the assumption of equal spacing, (5.13) can be simplified as follows by replacing t_i with $(i-1)t$:

$$Z_o = t\,v\,(i + \frac{2\,D_{x_{i,i-1}}}{D_{x_{i+1,i-1}} - 2\,D_{x_{i,i-1}}}). \tag{5.24}$$

Similarly,

$$Z_o = t\,v\,(i + \frac{2\,D_{y_{i,i-1}}}{D_{y_{i+1,i-1}} - 2\,D_{y_{i,i-1}}}), \tag{5.25}$$

and

$$Z_o = t\,v\,(i + \frac{2\,D_{i,i-1}}{D_{i+1,i-1} - 2\,D_{i,i-1}}). \tag{5.26}$$

5.3   Estimation of the Gabor Features

5.3.1   GABOR CORRELATION OPERATOR

The Gabor features are extracted by a Gabor correlation operator which consists of a set of Gabor functions. Gabor functions, originally introduced in the context of uncertainty theory for information theory[Gab46], have been widely used in image processing and machine vision [Dau88, PZ88, BLvdM89, ZG90, ZC91] since Daugman [Dau85] developed two-dimensional Gabor functions. Gabor functions consist of sinusoidal functions of time multiplied by a Gaussian function of time. A general form of the two-dimensional Gabor functions is expressed by [Dau88]

$$\begin{aligned} G(x,y|\alpha,\beta) \;=\; & \exp\{-\pi[(x-x_0)\alpha^2 + (y-y_0)\beta^2]\} \\ & \cdot \exp\{-2\pi i[u_0(x-x_0) + v_0(y-y_0)]\} \end{aligned} \tag{5.27}$$

where (x_0, y_0) are the position parameters which localize the function to a region in visual space, (u_0, v_0) are the modulation parameters which orient the function to a preferred direction and control the frequency space, (size of the perceptive field), and (α, β) are the scale parameters which determine spatial dimensions. The scale parameters (α, β) give one more degree of freedom for rotating the coordinate (x, y) out of the principal axes. For instance, $\alpha \neq \beta$ yields a cross term in xy. Varying the modulation and scale parameters in (5.27) generates whole families of Gabor functions.

To avoid a cross term in xy, we simply set

$$\alpha^2 = \beta^2 = \frac{2^{m-1}}{\pi M^2} \tag{5.28}$$

where M denotes the spatial dimension and m denotes the resolution level. By choosing the modulation parameters

$$u_0 = -\frac{2^{\frac{m}{2}-1}}{M}\cos(\frac{\pi n}{N}) \tag{5.29}$$

and

$$v_0 = -\frac{2^{\frac{m}{2}-1}}{M}\sin(\frac{\pi n}{N}), \tag{5.30}$$

we have

$$
\begin{aligned}
G(x,y|m,n) &= \exp\{\frac{i\pi 2^{\frac{m}{2}}}{M}[(x-x_0)\cos(\frac{\pi n}{N}) + (y-y_0)\sin(\frac{\pi n}{N})]\} \\
&\quad \exp\{-\frac{2^{m-1}}{M^2}[(x-x_0)^2 + (y-y_0)^2]\}
\end{aligned}
\tag{5.31}
$$

where N is the total number of orientations and n denotes a preferred orientation. Varying parameters m and n results in whole families of the Gabor functions. Figures 5.3 and 5.4 shows two Gabor functions with different m and n. The real and imaginary parts of the Gabor function are given in the left and right columns, respectively. Each column contains three plots representing different viewing directions. It is interesting to note that the real part is symmetric and the imaginary part is asymmetric.

5.3.2 COMPUTATIONAL CONSIDERATIONS

The Gabor correlation operator contains a set of two-dimensional multiresolution Gabor functions. The Gabor feature at resolution level m with orientation n is computed by

$$g(i,j|m,n)) = \int I(x,y)G(x-i,y-j|m,n)dx\,dy \tag{5.32}$$

where $I(x,y)$ is the image intensity function. Since $G(x,y|m,n)$ is a complex function, $g(i,j|m,n)$ is a complex number. For simplicity, we treat $g(i,j|m,n)$ as two real numbers. Let $g_r(i,j|m,n)$ and $g_i(i,j|m,n)$ denote the real and imaginary parts of $g(i,j|m,n)$. Gabor features are organized as

$$
\begin{aligned}
\underline{G}(i,j) &= [g_r(i,j|m_1,n_1), g_i(i,j|m_1,n_1), g_r(i,j|m_2,n_2), ..., \\
&\quad g_i(i,j|m_{k-1},n_{k-1}), g_r(i,j|m_k,n_k), g_i(i,j|m_k,n_k)]^T \tag{5.33}
\end{aligned}
$$

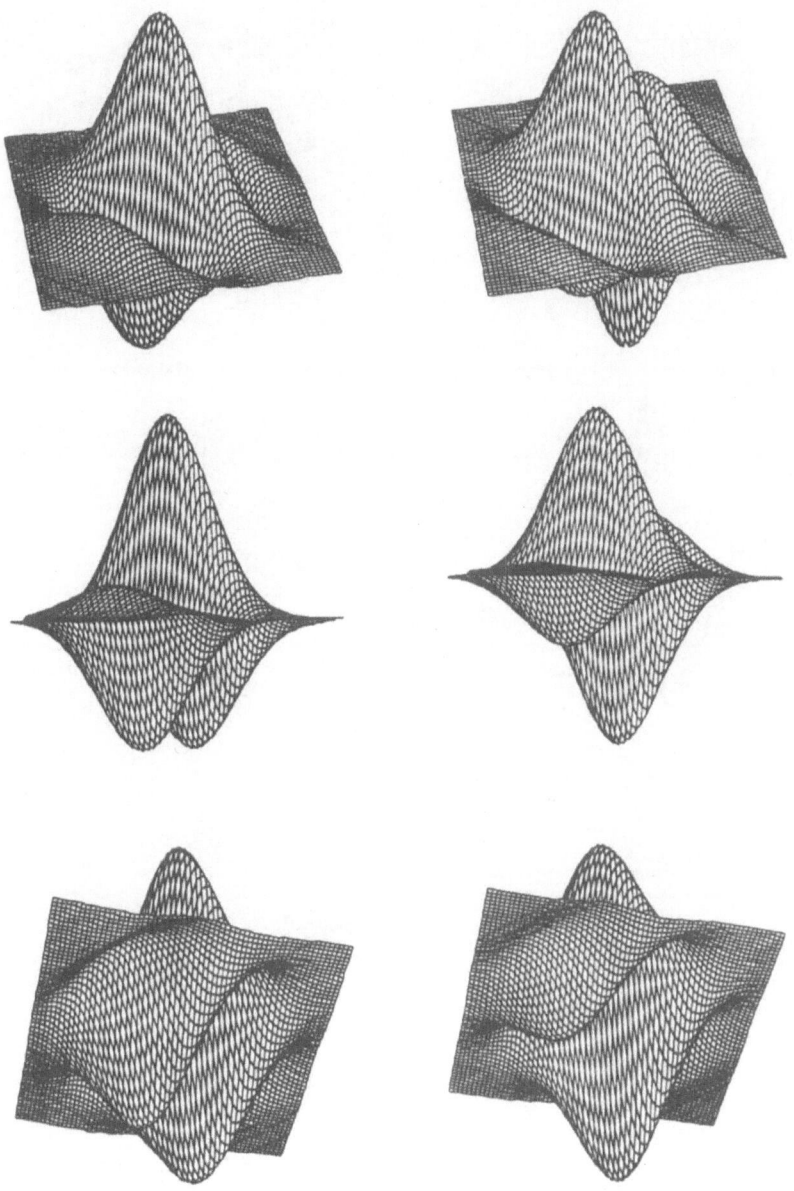

FIGURE 5.3. Perspective plots of the Gabor function for M=128, N=8, m=5 and n=2. Left column: real part. Right column: imaginary part.

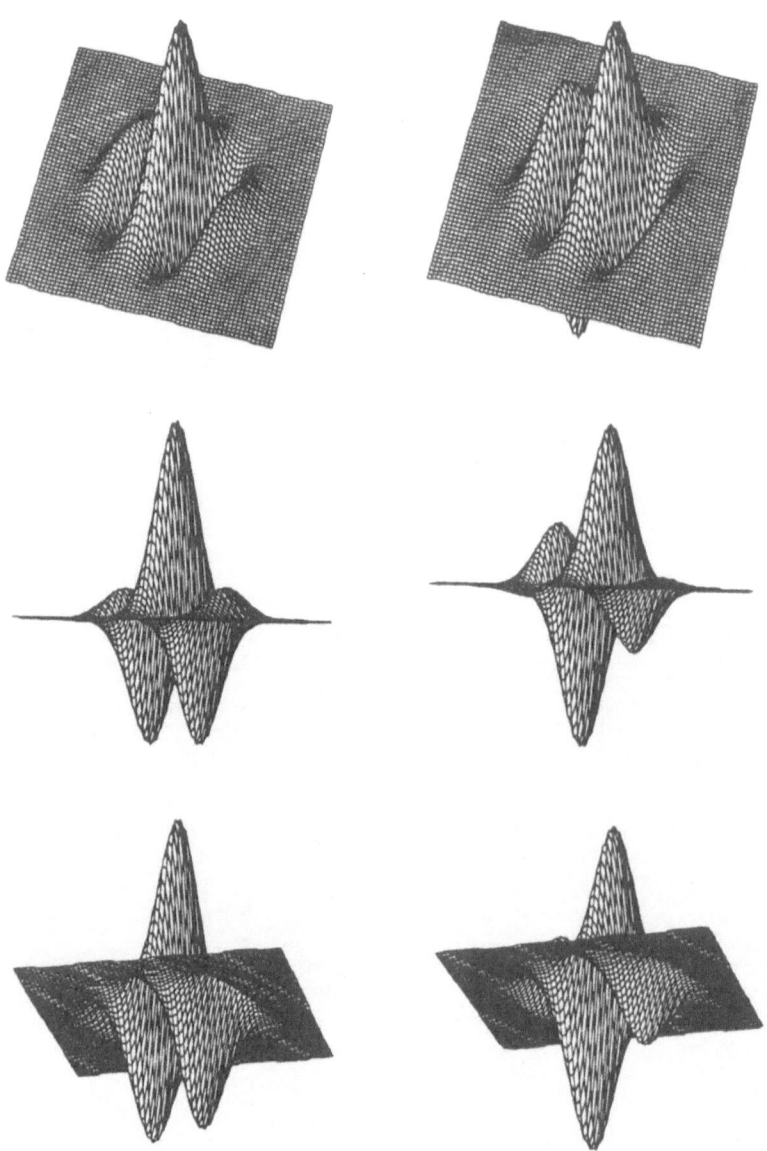

FIGURE 5.4. Perspective plots of the Gabor function for M=128, N=8, m=6 and n=6. Left column: real part. Right column: imaginary part.

where T is the transpose operator.

For implementation purposes, Gabor functions are discretized and truncated. Using a discrete Gabor function, (5.32) becomes

$$g(i, j|m, n)) = \sum_x \sum_y I(x, y)G(x - i, y - j|m, n). \qquad (5.34)$$

Since a large truncation error increases the sensitivity of the Gabor function to high frequency noise and lighting conditions, it is important to keep the truncation error as small as possible. From (5.31) it can be seen that the size (resolution level) of the perspective field is controlled by m for a fixed spatial dimension of M. In general, a small m corresponds to a large perspective field (low resolution). As shown in Figures 5.3 and 5.4, the first Gabor function has a large respective field ($m = 5$) and the second one has a small respective field ($m = 6$). Since the perspective field of the first Gabor function is too large to fit the window, it has been seriously truncated. In contrast, the second Gabor function fits the window well and no severe truncation has been observed. Hence, the truncation phenomenon is caused by forcing a low resolution Gabor function to fit within a small window. To reduce the truncation error, one might increase the spatial dimension M. However, a larger M dramatically increases computations.

As mentioned before, a Gabor function is weighted by a Gaussian function. Since the Gaussian function is bell shaped, the most of the area under the Gaussian function lies within six standard deviations of the mean. It is possible to keep the spatial frequency characteristics unchanged if we truncate the Gabor function to a central area with a diameter of six standard deviations. A general form of two-dimensional Gaussian function can be written as

$$f(x, y) = c \, \exp\{-\frac{[(x - x_0)^2 + (y - y_0)^2]}{2\sigma^2}\} \qquad (5.35)$$

where c is a scale. Comparing (5.35) with the Gaussian function in (5.31) gives the standard deviationindexStandard deviation

$$\sigma = M \, 2^{-\frac{m}{2}}, \qquad (5.36)$$

which is a function of M and m. Let D be the diameter of the truncated central area

$$D_k = k \, \sigma \qquad (5.37)$$

where k is a scale. The diameter of the truncated central area is determined by k and σ. The total area under the truncated Gaussian function can be estimated from the Gaussian probability table. For instance, if k is set at $2\sqrt{2}$, the diameter D_k is then $2\sqrt{2}\sigma = M \, 2^{\frac{3-m}{2}}$ which contains 84.2% of the area. If k is set at 6, then 99.8% of the area is covered by the truncated Gaussian function. Hence, it is safe to set k at 6. Tables 5.1 and 5.2 give the diameters of the truncated central area for $k = 2\sqrt{2}$ (84.2% of the area)

The error function for computing the disparity field based on two frames i and $(i+1)$ is defined as

$$
\begin{aligned}
E_{i,i+1} &= \sum_i \sum_j \sum_k \sum_l \|\underline{G}_i(i,j) - \underline{G}_{i+1}(i+k,j+l)\|^2 \, v_{i,j,k,l} \\
&+ \frac{A}{2} \sum_i \sum_j [\sum_k \sum_{r \in \Omega_{row}} (\sum_l (v_{i,j,k,l} - v_{i+r,j,k,l}))^2 \\
&+ \sum_l \sum_{c \in \Omega_{col}} (\sum_k (v_{i,j,k,l} - v_{i,j+c,k,l}))^2]
\end{aligned} \tag{5.42}
$$

where A is a constant, $\underline{G}_i(i,j)$ and $\underline{G}_{i+1}(i+k,j+l)$ are the Gabor feature vectors of the ith and $(i+1)$th images, respectively, and Ω_{row} is the row index set excluding $(0,0)$ for all neighbors in a $1 \times \omega$ window centered at point (i,j), and Ω_{col} is the column index set excluding $(0,0)$ for all neighbors in a $\omega \times 1$ window centered at point (i,j). If ω is set at 3, then Ω_{row} and Ω_{col} are $\{-1,1\}$. $\omega = 5$ gives $\{-2,-1,1,2\}$. The first term in (5.42) is to ensure that all points of two images are matched as closely as possible in a least squares sense based on the Gabor features. The second term is the combination of the row and column smoothness constraints on solution.

Expanding (5.42) gives

$$
\begin{aligned}
E_{i,i+1} &= \sum_i \sum_j \sum_k \sum_l \|\underline{G}_i(i,j) - \underline{G}_{i+1}(i+k,j+l)\|^2 \, v_{i,j,k,l} \\
&+ \frac{A}{2} \sum_i \sum_j \sum_k \sum_l [\sum_q \sum_{r \in \Omega_{row}} (v_{i,j,k,l} v_{i,j,k,q} \\
&+ v_{i+r,j,k,l} v_{i+r,j,k,q} - 2\, v_{i+r,j,k,l} v_{i,j,k,q}) \\
&+ \sum_p \sum_{c \in \Omega_{col}} (v_{i,j,k,l} v_{i,j,p,l} + v_{i,j+c,k,l} v_{i,j+c,p,l} \\
&- 2\, v_{i,j+c,k,l} v_{i,j,p,l})].
\end{aligned} \tag{5.43}
$$

On rearrangement of terms in the above equation, we have

$$
\begin{aligned}
E_{i,i+1} &= \sum_i \sum_j \sum_k \sum_l \|\underline{G}_i(i,j) - \underline{G}_{i+1}(i+k,j+l)\|^2 \, v_{i,j,k,l} \\
&+ \sum_p (W_o\, v_{i,j,k,l} v_{i,j,p,l} - \sum_{c \in \Omega'_{col}} W_c\, v_{i,j,k,l} v_{i,j+c,p,l})] \\
&+ \frac{A}{2} \sum_i \sum_j \sum_k \sum_l [\sum_q (W_o\, v_{i,j,k,l} v_{i,j,k,q} \\
&- \sum_{r \in \Omega'_{row}} W_r\, v_{i,j,k,l} v_{i+r,j,k,q})
\end{aligned} \tag{5.44}
$$

where Ω'_{row} and Ω'_{col} are the modified row and column index sets, and W_o, W_r and W_c are the weights. The weights and index sets can be determined as follows. If ω equals 3, then

$$W_o = 6, \quad W_x = -5|x| + 9, \quad x \in \Omega'_x$$

and

$$\Omega'_x = \{-2, -1, 1, 2\}, \quad \text{for } x = row, col.$$

If ω equals 5, then

$$W_o = 20, \quad W_x = \frac{10}{3}|x|^3 - 25|x|^2 + \frac{158}{3}|x| - 25, \quad x \in \Omega'_x$$

and

$$\Omega'_x = \{-4, -3, -2, -1, 1, 2, 3, 4\}, \quad \text{for } x = row, col.$$

Comparing (5.44) with the energy function

$$
E = -\frac{1}{2}\sum_i\sum_j\sum_k\sum_l\sum_m\sum_n\sum_p\sum_q T_{i,j,k,l;m,n,p,q}v_{i,j,k,l}v_{m,n,p,q}
$$
$$
-\sum_i\sum_j\sum_k\sum_l I_{i,j,kl} \tag{5.45}
$$

yields

$$
T_{i,j,k,l;m,n,p,q} = A\,[\delta_{j,n}\delta_{k,p}(\sum_{r\in\Omega'_{row}} W_r\,\delta_{i,m+r} - W_o\delta_{i,m})
$$
$$
+\delta_{i,m}\delta_{l,q}(\sum_{c\in\Omega'_{col}} W_c\,\delta_{j,n+c} - W_o\delta_{j,n})] \tag{5.46}
$$

and

$$I_{i,j,k,l} = -\|\underline{G}_i(i,j) - \underline{G}_{i+1}(i+k, j+l)\|^2 \tag{5.47}$$

where $\delta_{a,b}$ is the Dirac delta function. Therefore, computation of the disparity field can be carried out by neuron evaluation.

The initial state of the neurons is set as

$$
v_{i,j,k,l} = \begin{cases} 1 & \text{if } I_{i,j,k,l} = \max_{p,q}(I_{i,j,p,q}) \\ 0 & \text{otherwise} \end{cases} \tag{5.48}
$$

where $I_{i,j,k,l}$ is the bias input. As with static stereo, the minimal mapping theory is used for handling the case of two neurons having the same largest inputs, and a deterministic decision rule is used for updating neuron states.

5.5   Experimental Results

The algorithm has been tested on a number of real image sequences. The total test data covers the forward and backward motions with both small and large motions. To achieve subpixel accuracy, the Chebyshev polynomial interpolation technique presented in Chapter 4 was used for locating the conjugate points in the small motion case. Unlike lateral motion stereo, longitudinal motion stereo needs the two-dimensional Chebyshev polynomials to interpolate the image in both horizontal and vertical directions. Details about the two-dimensional Chebyshev polynomials are given in Chapter 6. For testing purposes, we randomly picked up three successive frames from each image sequence. To avoid the accumulated error, we used the $(i+1)$th and $(i-1)$th image frames to estimate $D_{x_{i+1,i-1}}$. Two disparity fields were generated based on the estimated disparity values. Figure 5.5 shows the first three frames from an Apple sequence of four frames. The image size is 196×196. The image coordinate system is centered at the FOE, thus at the center of the image. A total of 40 Gabor functions (five resolutions times eight orientations) were used to extract image features. No alignment was made and the maximum disparity was measured by hand. The disparity maps were superimposed on the image as shown in Figure 5.6. In order to keep the original image visible, only a few disparity vectors were plotted. Since it is straightforward to recover a depth map from the disparity fields, commonly only the disparity values are computed, unless a depth map is required.

5.6   Discussion

We have presented a longitudinal motion stereo algorithm based on three frames. Gabor features were used as measurement primitives for matching. Experimental results show that Gabor functions capture the object features very efficiently and the disparity values computed by the algorithm based on the Gabor feature vectors are reliable and accurate. Recently, we have used Gabor functions for detecting targets in infrared imagery [ZC91]. The preliminary results are very encouraging. The detection algorithm is insensitive to variations in target contrast, size, shape and orientation. However, one major problem with Gabor functions is a high computational complexity. To reduce the computational complexity, we can use the fast Fourier transform to do convolution [BCG90]. By converting the image data from the spatial domain into the frequency domain, the convolution operation can be carried out by multiplication. Although three frames are sufficient to recover the depth information, it is highly desirable to develop multiple frame-based (more than three frames) algorithms for longitudinal motion stereo in order to achieve high accuracy and robustness.

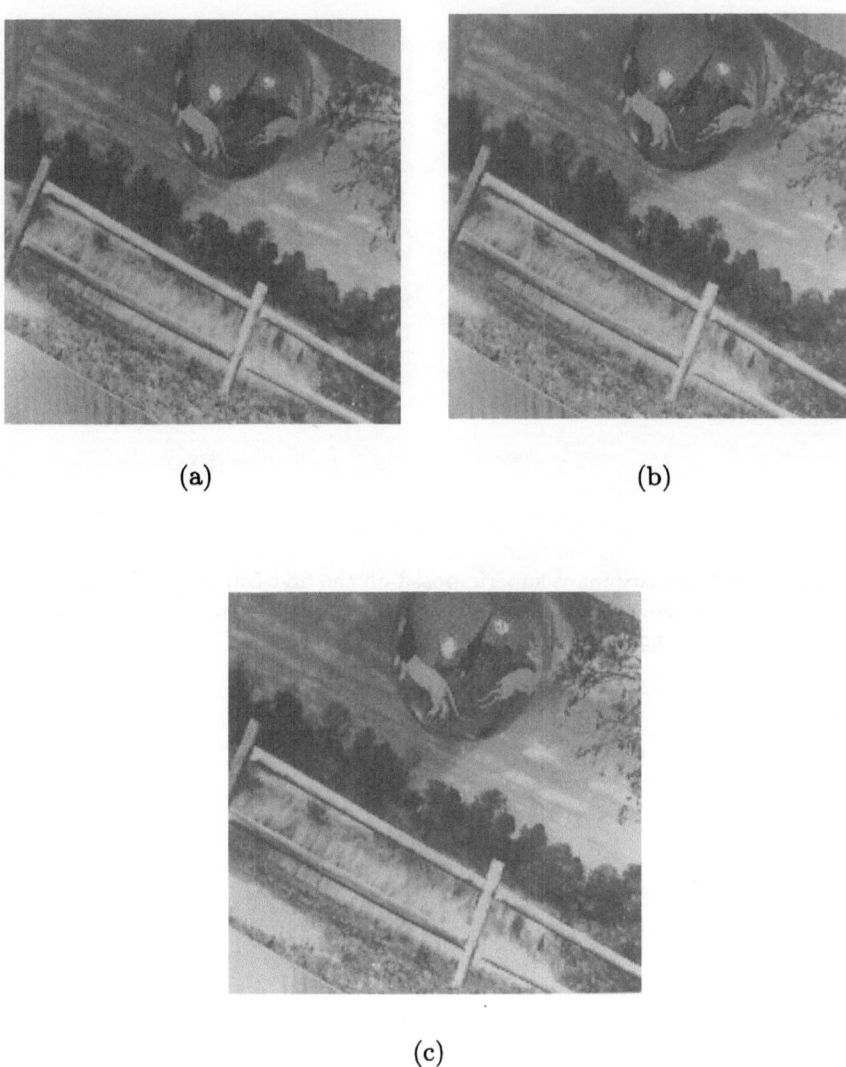

FIGURE 5.5. The Apple motion sequence. (a) The first frame. (b) The second frame. (c) The third frame.

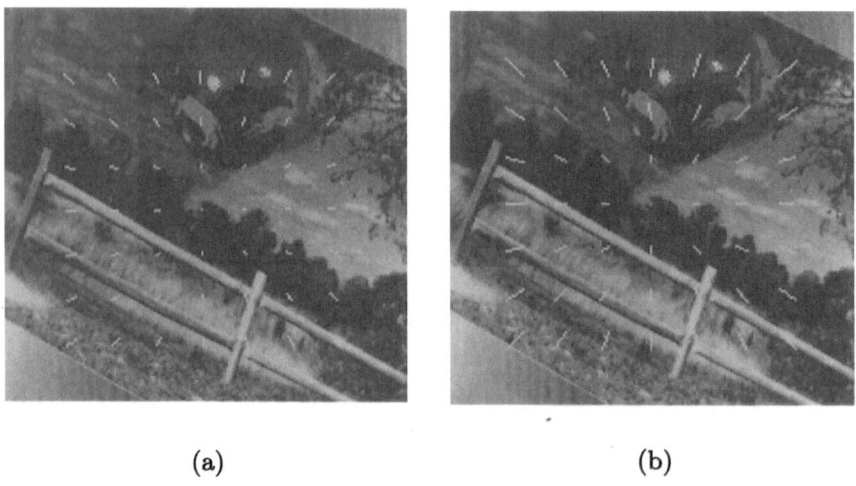

(a)   (b)

FIGURE 5.6. Disparity maps superimposed on the first frame. (a) Disparity values estimated by using the first and second frames. (b) Disparity values estimated by using the first and third frames.

6

Computation of Optical Flow

6.1 Introduction

Optical flow is the distribution of apparent velocities of moving brightness patterns in an image. Ideally the optical flow corresponds to the motion field, but this is not always true [Hor86]. It is common to assume that the optical flow is not too different from the motion field. Under this assumption, the optical flow can be used for segmenting images into regions and estimating the object motion in the scene [Adi85].

Figure 6.1 shows a camera configuration used in the motion estimation. As an object is moving in front of a camera, the corresponding brightness pattern in the image moves too. At each instant in time, the velocity vector associated with each point on the object projects onto the image plane as a motion vector as shown in Figure 6.2. Let $u(i,j)$ and $v(i,j)$ be the i and j components of the optical flow vector at the point (i,j). If the brightness of a pattern varies smoothly with i, j, and time t, then the optical flow vector can be defined as

$$u(i,j) = \frac{di}{dt} \qquad (6.1)$$

and

$$v(i,j) = \frac{dj}{dt}. \qquad (6.2)$$

For implementation purposes, the differential equations (6.1) and (6.2) are commonly replaced with difference equations

$$u(i,j) \approx \frac{i_{t+h} - i_t}{h} \qquad (6.3)$$

and

$$v(i,j) \approx \frac{j_{t+h} - j_t}{h} \qquad (6.4)$$

where h denotes the time interval, and (i_t, j_t) and (i_{t+h}, j_{t+h}) are positions of the brightness pattern at time t and $(t+h)$, respectively. Computation of optical flow is then simplified so as to find the displacements

$$\Delta i = (i_{t+h} - i_t) \quad \text{and} \quad \Delta j = (j_{t+h} - j_t)$$

of the brightness pattern.

Existing approaches to computation of optical flow can be divided into two types: image intensity-based or token-based approaches. The intensity-based approach basically relies on the assumption that the changes in intensity are strictly due to the motion of the object and uses the image

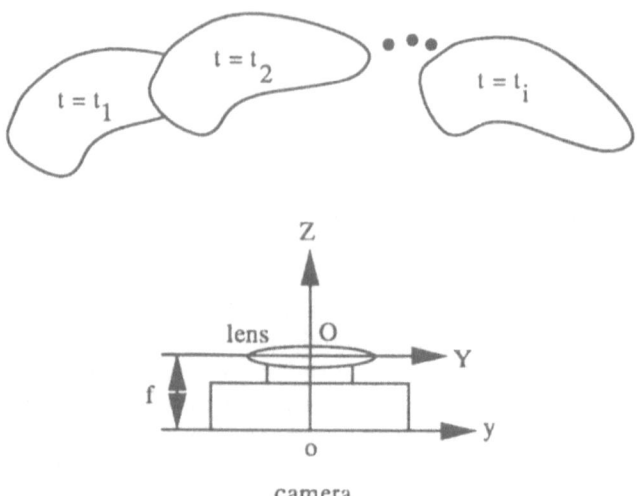

FIGURE 6.1. A object is moving in front of a camera.

intensity values and their spatial and temporal derivatives to compute the optical flow. Limb and Murphy [LM75] utilize the relation between spatial and temporal differential signals to estimate the displacement of an object in a television scene under the assumption that the displacement is constant within a block of image pixels. In order to make the constant displacement assumption more realistic to situations in which there are multiple moving objects, occluding objects, and different parts of the same object moving with different displacements, Netravali and Robbins [NR79] proposed a recursive algorithm for estimating the displacements of multiple moving objects in a television scene by minimizing a functional of the prediction error. Since the update at each pixel is based on gradient information at a single point, the method is more sensitive to noise than Limb's block approach. Subsequently, they extended their method by using the gradient information at a number of pixels in the neighborhood of the point [NR80]. Fennema and Thompson [FT79] developed a nonmatching method for computing the speed and direction of the one or more moving objects in a scene using a clustering technique based on spatial and temporal derivatives of the intensity. To determine the actual velocity, a modification of the Hough Transform was used. All these methods are applicable to rigid body translation only and give a dense flow field. Expanding the intensity function as a first order Taylor series, Horn and Schunck [HS81] derived an optical flow equation using brightness constancy and spatial smoothness constraints. An iterative method for solving the resulting equation was de-

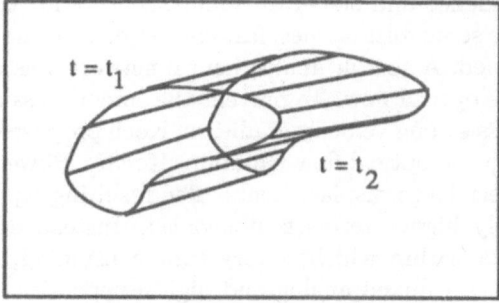

FIGURE 6.2. Motion vectors.

veloped. This method usually fails to detect both discontinuous locations of the velocity field due to oversmoothing and large displacements due to first order approximation.

Recently, Gennert and Negahdaripour [GN87] have proposed to replace the brightness constancy constraint in the optical flow equation with a more general constraint, which permits a linear transformation between the image intensity values. Instead of generating a dense flow field, Hildreth [Hil83] proposed a method based on the intensity gradients to estimate the velocities only along the zero-crossing contours. Nagel [Nag83] also investigated an approach to track corners and estimate the displacement at the corners by using partial derivatives of the intensity function.

The token-based approach is to consider the motion of tokens such as edges, corners and linear features in an image. The main reason to consider a token-based approach is that the tokens are less sensitive to some of the difficulties associated with variations of the image intensity. Thompson and Barnard [Tho81] developed a probabilistic relaxation labeling scheme for computing displacements based on corners, spots, and similar structures in images. Under the assumption of smooth, rigid body motion and nonspare distribution of feature points on a surface, Prager and Arbib [PA83] developed a MATCH algorithm for computing the optical flow. However, the token-based approaches give information about object motion and shape only at edges, corners and linear features. An interpolation procedure has to be included when dense flow field is required.

Recently, several researchers have used a neural network for computing optical flow based on either intensity values or tokens [GL87, Koc87]. In [GL87], a neural network proposed by Hopfield and Tank [HT85] was used to realize the Ullman's Minimal Mapping Theory [Ull79] for computing the optical flow based on feature points of a nonrigid moving object. They also used the same neural network to implement the structural theory for solving correspondence and structure simultaneously for the rigid motion problems. Despite some mismatches, fast convergence of the neural network was always obtained. A similar analog neural network was used in [Koc87] for computing the optical flow. To prevent the smoothness constraint from taking effect across strong velocity gradients, Koch proposed to incorporate a line process into the optical flow equation [Koc87]. However, no attempt was made to detect large displacements. The resulting equation involving cubic and possibly higher terms is nonconvex. Instead of using a simulated annealing algorithm which is very time consuming, a deterministic algorithm based on a mixed analog and digital network is used to obtain a suboptimal solution. As reported in [HKLM88], convergence of such a network was obtained within a couple of analog-digital cycles. Basically, the analog-digital network approach is the first to use Horn's optical flow equation to find to find a smoothest solution, and then to update the line process by lowering the energy function of the network repeatedly. Several impressive examples using synthetic and video images were presented in [HKLM88].

In order to obtain a dense optical flow field, it seems that the intensity-based approach is preferable. However, the intensity value may be corrupted by noise and also their partial derivatives are rotation variant. It is difficult to detect rotating objects in natural images, based on such measurement primitives. Under the assumption that the changes in intensity are strictly due to object motion, we may use rotation invariant principal curvatures of the intensity function to compute the optical flow. In this chapter, we use a neural network with a maximum evolution function to compute optical flow based on the intensity values and on the principal curvatures under the local rigid motion assumption and smoothness constraints. For detecting motion discontinuities, a line process is commonly used [Koc87] for locating motion discontinuities. However, without exactly knowing the occluding elements, the detected discontinuities may be shifted. In order to detect discontinuities accurately, we first detect the occluding elements from the initial motion measurements, then use a line process to locate the discontinuities by using the information about the detected occluding elements. The intensity values and their principal curvatures are estimated with subpixel accuracy by using a polynomial fitting technique. When a large window is applied, smooth estimates can be obtained from noisy observations. To ensure convergence of the network, deterministic decision rules are used. Since the neurons and lines are updated in parallel and the network is locally connected, this algorithm can be implemented in parallel and is well suited for VLSI implementation. To improve the accuracy, our algorithm also uses multiple frames to compute optical flow. Natural images are often degraded by the imaging system. Based on such imperfect observations, it is difficult to compute optical flow accurately, especially near motion discontinuities. To improve the accuracy of the solution, we use multiple frames. Two algorithms, batch and recursive, are presented.

The batch algorithm simultaneously integrates information from all images by embedding them into the bias inputs of the network, while the recursive algorithm uses a recursive least squares (RLS) method to update the bias inputs of the network. Both of these methods need to compute the optical flow no more than twice. Hence, fewer computations are needed and the recursive algorithm is amenable for real time applications. The performance of all these methods is illustrated by using synthetic and natural images.

6.2 Estimation of Intensity Values and Principal Curvatures

In order to reduce the effects of noise and quantization error, a bivariate polynomial fitting technique is used for the estimation of intensity values and their principal curvatures from discrete observations. The estimation problem can then be formulated to find polynomial coefficients such that the square error

$$\epsilon_{i,j} = \sum_{x \in \Omega} \sum_{y \in \Omega} (\tilde{g}(i+x, j+y) - \hat{g}(i+x, j+y))^2 \tag{6.5}$$

between the estimate $\hat{g}(i,j)$ and the observation $\tilde{g}(i,j)$ is minimized. In (6.5), Ω is an index set $\{-\omega, -\omega+1, ..., \omega-1, \omega\}$.

6.2.1 ESTIMATION OF POLYNOMIAL COEFFICIENTS

Bivariate discrete Chebyshev polynomials can be constructed by using the tensor product technique. We assume that in each neighborhood of an image point (i, j) the underlying intensity function can be approximated by a fourth order polynomial. We represent the intensity function in a window of size $2\omega + 1$ by $2\omega + 1$ by

$$\hat{g}(i+x, j+y) = \underline{\mathbf{CH}}^t(x) \underline{\mathbf{Q}} \underline{\mathbf{CH}}(y) \tag{6.6}$$

where $\hat{g}(i+x, j+y)$ is the approximated continuous intensity function, t denotes the transpose operator,

$$\underline{\mathbf{CH}}^t(x) = [Ch_0(x) \ Ch_1(x) \ Ch_2(x) \ Ch_3(x) \ Ch_4(x)]$$

and

$$\underline{\mathbf{CH}}^t(y) = [Ch_0(y) \ Ch_1(y) \ Ch_2(y) \ Ch_3(y) \ Ch_4(y)]$$

are polynomial vectors, and

$$\underline{\mathbf{Q}} = \begin{bmatrix} a_{0,0} & a_{0,1} & a_{0,2} & a_{0,3} & a_{0,4} \\ a_{1,0} & a_{1,1} & a_{1,2} & a_{1,3} & a_{1,4} \\ a_{2,0} & a_{2,1} & a_{2,2} & a_{2,3} & a_{2,4} \\ a_{3,0} & a_{3,1} & a_{3,2} & a_{3,3} & a_{3,4} \\ a_{4,0} & a_{4,1} & a_{4,2} & a_{4,3} & a_{4,4} \end{bmatrix}$$

is the coefficient matrix. By minimizing the square error (6.5) and taking advantage of the orthogonality of the polynomial set, the coefficients $\{a_{m,n}\}$ are obtained as

$$a_{m,n} = \frac{\sum_{y\in\Omega} \sum_{x\in\Omega} Ch_m(x)\, Ch_n(y)\, g(i+x, j+y)}{\sum_{u\in\Omega} \sum_{v\in\Omega} Ch_m^2(u)\, Ch_n^2(v)} \qquad (6.7)$$

where $\{g(i+x, j+y)\}$ are the observed intensity values. Once the coefficients are calculated, the subpixel intensity values can be estimated from (6.6). A smooth intensity function can be obtained if low order polynomials and large windows are used.

6.2.2   COMPUTING PRINCIPAL CURVATURES

The principal curvatures at (i, j) are defined as the maximum and minimum values of the normal curvatures of the intensity function at that point [O'N66]. A notable property of principal curvatures is that they are rotation invariant, which is useful for detecting rotating objects. The principal curvatures can be expressed in terms of Gaussian and mean curvatures of intensity function as follows [O'N66]:

$$k_1(i, j) = \mathbf{M} + (\mathbf{M}^2 - \mathbf{G})^{\frac{1}{2}} \qquad (6.8)$$

and

$$k_2(i, j) = \mathbf{M} - (\mathbf{M}^2 - \mathbf{G})^{\frac{1}{2}} \qquad (6.9)$$

where $k_1(i, j)$ and $k_2(i, j)$ are the principal curvatures, $\mathbf{G}$ and $\mathbf{M}$ are the Gaussian and mean curvatures defined as

$$\mathbf{G} = \frac{\partial^2 g(i, j)}{\partial i^2}\, \frac{\partial^2 g(i, j)}{\partial j^2} - (\frac{\partial^2 g(i, j)}{\partial i \partial j})^2$$

and

$$\mathbf{M} = \frac{1}{2}(\frac{\partial^2 g(i, j)}{\partial i^2} + \frac{\partial^2 g(i, j)}{\partial j^2}),$$

respectively, under the assumption that

$$\frac{\partial^2 g(i, j)}{\partial i \partial j} = \frac{\partial^2 g(i, j)}{\partial j \partial i}.$$

Using the estimated continuous intensity function, the second order partial derivatives of the intensity function can be calculated with subpixel accuracy as

$$\frac{\partial^2 g(i, j)}{\partial i^2}\Big|_{i=i+x, j=j+y} = \frac{\partial^2 \hat{g}(i+x, j+y)}{\partial x^2}$$

$$= \frac{d^2}{dx^2}(\mathbf{CH}^t(x))\, \underline{\mathbf{Q}}\, \underline{\mathbf{CH}}(y) \qquad (6.10)$$

$$\frac{\partial^2 g(i,j)}{\partial i \partial j}\Big|_{i=i+x,j=j+y} = \frac{\partial^2 \hat{g}(i+x,j+y)}{\partial x \partial y}$$

$$= \frac{d}{dx}(\underline{\mathbf{CH}}^t(x)) \ \underline{\mathbf{Q}} \ \frac{d}{dy}(\underline{\mathbf{CH}}(y)) \qquad (6.11)$$

$$\frac{\partial^2 g(i,j)}{\partial j^2}\Big|_{i=i+x,j=j+y} = \frac{\partial^2 \hat{g}(i+x,j+y)}{\partial y^2}$$

$$= \underline{\mathbf{CH}}^t(x) \ \underline{\mathbf{Q}} \ \frac{d^2}{dy^2}(\underline{\mathbf{CH}}(y)) \qquad (6.12)$$

$$for \quad -0.5 \le x, y < 0.5.$$

For simplicity of notation, we use $g_{xx}(x,y)$, $g_{xy}(x,y)$, and $g_{yy}(x,y)$ to represent the second order partial derivatives of the subpixel intensity function. After substitution of the second order partial derivatives, $\mathbf{M}$, (6.8) and (6.9) become

$$k_1(i+x,j+y) = \frac{1}{2}\{[(g_{xx}(x,y) - g_{yy}(x,y))^2 + 4g_{xy}^2(x,y)]^{\frac{1}{2}}\}$$

$$+ g_{xx}(x,y) + g_{yy}(x,y)\} \qquad (6.13)$$

and

$$k_2(i+x,j+y) = g_{xx}(x,y) + g_{yy}(x,y) - k_1(i+x,j+y) \qquad (6.14)$$

$$for \quad -0.5 \le x, y < 0.5.$$

Therefore, the principal curvatures of the subpixel intensity function can be obtained from the subpixel partial derivatives directly.

6.2.3 ANALYSIS OF FILTERS

Rewrite the coefficient matrix as

$$\underline{\mathbf{Q}} = \sum_{x' \in \Omega} \sum_{y' \in \Omega} \underline{\mathbf{Q}}(x',y') \ g(i+x',j+y') \qquad (6.15)$$

where $\mathbf{Q}(x',y')$ is a matrix with the (m,n)th element

$$\frac{Ch_m(x') \ Ch_n(y')}{\sum_{u \in \Omega} \sum_{v \in \Omega} Ch_m^2(u) \ Ch_n^2(v)}.$$

By replacing the matrix $\underline{\mathbf{Q}}$ in (6.6) with (6.15) we have

$$\hat{g}(i+x,j+y) = \sum_{x' \in \Omega} \sum_{y' \in \Omega} F(x,y;x',y') \ g(i+x',j+y') \qquad (6.16)$$

where $F(x, y; x', y')$ is determined by

$$F(x, y; x', y') = \underline{CH}^t(x) \ \underline{Q}(x', y') \ \underline{CH}(y). \tag{6.17}$$

Similarly, (6.10), (6.11) and (6.12) can be written as

$$\frac{\partial^2 g(i, j)}{\partial i^2}|_{i=i+x, j=j+y} = \sum_{x' \in \Omega} \sum_{y' \in \Omega} F_{xx}(x, y; x', y') g(i + x', j + y') \tag{6.18}$$

$$\frac{\partial^2 g(i, j)}{\partial i \partial j}|_{i=i+x, j=j+y} = \sum_{x' \in \Omega} \sum_{y' \in \Omega} F_{xy}(x, y; x', y') g(i + x', j + y') \tag{6.19}$$

$$\frac{\partial^2 g(i, j)}{\partial j^2}|_{i=i+x, j=j+y} = \sum_{x' \in \Omega} \sum_{y' \in \Omega} F_{yy}(x, y; x', y') g(i + x', j + y') \tag{6.20}$$

where

$$F_{xx}(x, y; x', y') = \frac{d^2}{dx^2}(\underline{CH}^t(x)) \ \underline{Q} \ \underline{CH}(y) \tag{6.21}$$

$$F_{xy}(x, y; x', y') = \frac{d}{dx}(\underline{CH}^t(x)) \ \underline{Q} \ \frac{d}{dy}(\underline{CH}(y)) \tag{6.22}$$

$$F_{yy}(x, y; x', y') = \underline{CH}^t(x) \ \underline{Q} \ \frac{d^2}{dy^2}(\underline{CH}(y)) \tag{6.23}$$

are the corresponding filters.

It is assumed that the image is corrupted by additive white noise with zero mean and variance σ_n^2

$$\tilde{g}(i, j) = g(i, j) + n(i, j) \tag{6.24}$$

where $\tilde{g}(i, j)$ and $g(i, j)$ are the observed and original intensity functions, respectively. The expected value of the output of the filter $F(x, y; x', y')$ can be

$$\mathbf{E}\{\sum_{x' \in \Omega} \sum_{y' \in \Omega} F(x, y; x', y') \ \tilde{g}(i + x', j + y')\}$$

$$= \sum_{x' \in \Omega} \sum_{y' \in \Omega} F(x, y; x', y') \ g(i + x', j + y'). \tag{6.25}$$

Accordingly, the variance is given by

$$\mathbf{Var}(\sum_{x' \in \Omega} \sum_{y' \in \Omega} F(x, y; x', y') \tilde{g}(i + x', j + y'))$$

$$= \mathbf{E}\{(\sum_{x' \in \Omega} \sum_{y' \in \Omega} F(x, y; x', y') \ n(i + x', j + y'))^2\}$$

$$= \sigma_n^2 \sum_{x' \in \Omega} \sum_{y' \in \Omega} F^2(x, y; x', y'). \tag{6.26}$$

By using (6.17) and noting the orthogonality of the polynomial set, it is straightforward to show that

$$\sum_{x' \in \Omega} \sum_{y' \in \Omega} F^2(x, y; x', y')$$

$$= \sum_{x' \in \Omega} \sum_{y' \in \Omega} (\underline{CH}^t(x) \, \underline{Q}(x', y') \, \underline{CH}(y))^2$$

$$= \sum_{m=0}^{4} \frac{Ch_m^2(x)}{\sum_{u \in \Omega} Ch_m^2(u)} \sum_{n=0}^{4} \frac{Ch_n^2(y)}{\sum_{v \in \Omega} Ch_n^2(v)}. \tag{6.27}$$

Hence, the variance of the filter output is

$$\mathbf{Var}(\sum_{x' \in \Omega} \sum_{y' \in \Omega} F(x, y; x', y') \tilde{g}(i + x', j + y'))$$

$$= \sigma_n^2 \sum_{m=0}^{4} \frac{Ch_m^2(x)}{\sum_{u \in \Omega} Ch_m^2(u)} \sum_{n=0}^{4} \frac{Ch_n^2(y)}{\sum_{v \in \Omega} Ch_n^2(v)}. \tag{6.28}$$

Similarly, beginning with (6.21), (6.22) and (6.23), one can show that the variances of the outputs of the filters $F_{xx}(x, y; x', y')$, $F_{xy}(x, y; x', y')$, and $F_{yy}(x, y; x', y')$ are

$$\mathbf{Var}(\sum_{x' \in \Omega} \sum_{y' \in \Omega} F_{xx}(x, y; x', y') \tilde{g}(i + x', j + y'))$$

$$= \sigma_n^2 \sum_{m=2}^{4} \frac{(\frac{d^2}{dx^2} Ch_m(x))^2}{\sum_{u \in \Omega} Ch_m^2(u)} \sum_{n=0}^{4} \frac{Ch_n^2(y)}{\sum_{v \in \Omega} Ch_n^2(v)}$$

$$\mathbf{Var}(\sum_{x' \in \Omega} \sum_{y' \in \Omega} F_{xy}(x, y; x', y') \tilde{g}(i + x', j + y'))$$

$$= \sigma_n^2 \sum_{m=1}^{4} \frac{(\frac{d}{dx} Ch_m(x))^2}{\sum_{u \in \Omega} Ch_m^2(u)} \sum_{n=1}^{4} \frac{(\frac{d}{dy} Ch_n(y))^2}{\sum_{v \in \Omega} Ch_n^2(v)} \tag{6.29}$$

$$\mathbf{Var}(\sum_{x' \in \Omega} \sum_{y' \in \Omega} F_{yy}(x, y; x', y') \tilde{g}(i + x', j + y'))$$

$$= \sigma_n^2 \sum_{m=0}^{4} \frac{Ch_m^2(x)}{\sum_{u \in \Omega} Ch_m^2(u)} \sum_{n=2}^{4} \frac{(\frac{d^2}{dy^2} Ch_n(y))^2}{\sum_{v \in \Omega} Ch_n^2(v)}.$$

It is clear from the above equations that the variances become smaller and smaller as the window size increases. For instance, when the window size is five, the variance of the output of the filter $F(x, y; x', y')$ at $(x, y) = (0, 0)$ is σ_n^2. When the window size is 11, the variance is significantly reduced to

$0.044\sigma^2$. As pointed out in Chapter 3, a large window causes some loss of local information due to smoothing which smears or erases local features. If one desires to retain the local features, then a small window may be used, but more noise remains and the estimated intensity function, their derivatives, and hence their principal curvatures are less accurate.

6.3   Neural Network Formulation

6.3.1   PHYSIOLOGICAL CONSIDERATIONS

Our neural approach has been motivated by the landmark discoveries made by Hubel and Wiesel [HW65, HW77] about the brain mechanisms of vision. The classical microelectrode studies done by Hubel and Wiesel in cats and monkeys show that the visual cortex, a few millimeters thick neuronal tissue, is organized in a topographic, laminar and columnar fashion [HW65, HW77]. The image on the retina is first projected to the lateral geniculate bodies and then from there to the visual cortex in a strict topographical manner. The neurons in the visual cortex are arranged in layers and grouped according to several stimulus parameters such as eye dominance, receptive field axis orientation and receptive field position. The groupings take the form of vertically arranged parallel slabs spanning the full cortical thickness. The optical nerve fibers arriving from the lateral geniculate bodies mostly terminate in layer 4 of visual area 17, yielding a cortical representation of the retina. From area 17 the visual signals pass to adjacent area 18 and other higher visual areas such as the middle temporal (MT), each with a complete topographic map of the visual field [TTA+81, BPNA81, CWK82].

Figure 6.3 is an idealized and speculative scheme for the hypercolumns of the visual area 17 [HW77]. Neurons with similarly orientation- and direction-selectivities are stacked in columns which are perpendicular to the cortical surface. For simplicity, blobs and ocular dominance columns are not included. All the neurons (simple, complexNeurons, complex, and hypercomplex) within a column have the same receptive field axis orientation. For instance, if an electrode is penetrated into the cortex in a direction perpendicular to the cortex surface, all the neurons encountered show the same axis orientation. If the electrode goes in a direction parallel to the cortex surface, there occurs a regular shift in the axis orientation, approximately five to ten degrees for every advance of $25 - 50$ μm. Over a distance of approximately one millimeter, there is roughly a full cycle of rotation $(180°)$. A set of orientation columns representing a full rotation of $180°$ together with an intersecting pair of ocular dominance columns forms a hypercolumn. Each hypercolumn as an elementary unit of the visual cortex is responsible for a certain small area of the visual field and encodes a complete feature description of the area by the activity of neurons. Ad-

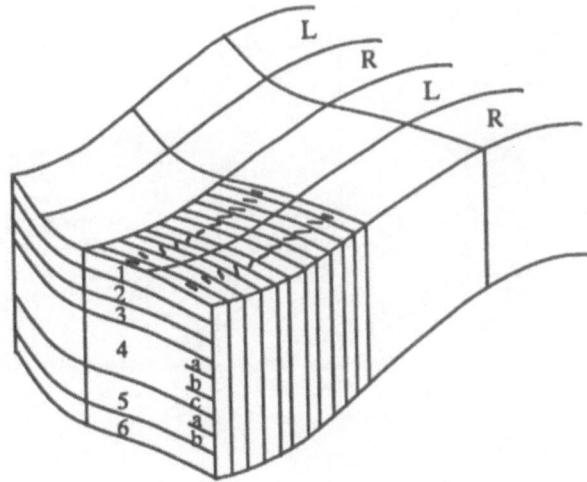

FIGURE 6.3. An idealized and speculative scheme for the hypercolumns of visual area 17. The blocks denoted by thick lines represent hypercolumns containing complete sets of orientation columns. The thin lines separate individual orientation columns. **L** and **R** denote the areas corresponding to left and right eyes. (Reprinted with permission from [ZC90] Figure 1, ©1990 IEEE.)

vancing more than one millimeter produces a displacement in the visual field, out of the area where one started and into an entirely new area. The simple neuron is orientation-selective and the complex neuron is direction-selective. The simple neuron responds best to a stationary line which is oriented with the axis of the receptive field. For the complex neurons, not only the orientation of the line but also the stimulus speed and motion direction are important. An oriented line produces strong responses to a complex neuron if it moves at an optimal speed in a direction perpendicular to the receptive field axis within the receptive field. Approximately half of the complex neurons respond only to one direction of movement. If the speed is less or greater than the optimum, the neuron's firing frequency tends to fall off sharply. The optimal speed varies from neuron to neuron. For instance, in cats it varies approximately from 0.1°/sec to 20°/sec [HW65]. Hence, the complex neurons are direction- and speed-selective, or velocity-selective. We assume that the complex neurons within a column can be further grouped according to their speed selectivity. Figure 6.4 shows a possible two-dimensional grouping pattern of the complex neurons within a hypercolumn. Each circle represents one or more neurons since several neurons may have the same velocity selectivity. The coordinates of the circle indicate the velocity selectivity of the neurons.

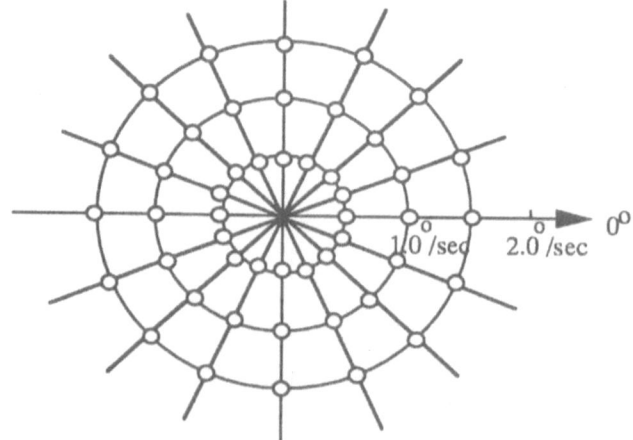

FIGURE 6.4. A two-dimensional grouping pattern for the neurons within a hypercolumn. Neurons are arranged according to their direction and speed selectivity. Each circle denotes one or more neurons.

The MT region is also a "motion area." Neurons in the MT area are predominantly direction-selective, with approximately 90% showing some direction selectivity and 80% showing high selectivity, and are arranged in columns according to direction selectivity [Mon80, CWK82, ADG84]. Area 17 projects to the MT area in a very unique way: (1) the projection only happens between columns with similar directionality; (2) neurons projecting from a given location in area 17 diverge to several periodically spaced locations in the MT area, and several locations in area 17 converge upon a given location in the MT area. These properties probably play very important roles in maintaining axis and direction selectivity and forcing the neighboring receptive fields to have the same directional preference. Figure 6.5 shows such a projection pattern.

6.3.2 COMPUTATIONAL CONSIDERATIONS

For implementation purposes, we assume that the neurons in a hypercolumn are uniformly distributed over a two-dimensional Cartesian plane, as shown in Figure 6.6, since images are uniformly sampled and computing optical flow means finding the conjugate points in images. The conjugate point can be found by checking every image pixel within a neighborhood in the successive frame based on the measurement primitives. The maximum search range, which is the maximum displacement, can be determined by the maximum optimal speed. To improve the accuracy of the solution, the

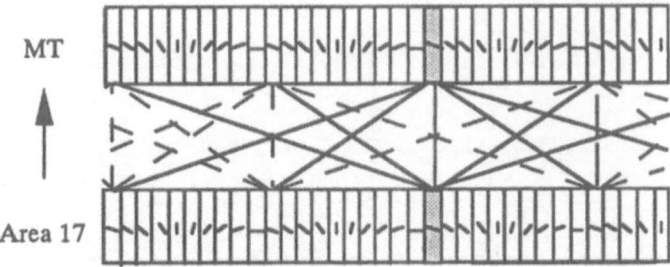

FIGURE 6.5. Projection pattern from area 17 to the MT area. For simplicity, only the cross-section of the four different orientation selective hypercolumns are shown for each area. The actual arrangements are more complex.

velocity component ranges can be further sampled using bins of size W, where W is a real number less than 1.

We assume that each hypercolumn represents a single image pixel or subpixel (if the image is subsampled). If the maximum displacement is D, then about $(2D + 1)^2$ mutually exclusive neurons are needed for each pixel and a total number of $N_r \times N_c \times (2D + 1)^2$ neurons are required for an $N_r \times N_c$ image. Since two objects cannot occupy the same place at the same time, only one velocity value can be assigned to each pixel. Therefore, in each hypercolumn, only one neuron is in the active state. The velocity value can be determined according to its direction selectivity. Figure 6.7 shows such a network with the small frames for the hypercolumns and the circles for the neurons. In fact, each small frame contains many neurons. For simplicity, only a few neurons are present in each frame. Each neuron receives a bias input from the outside world. The bias input may consist of several different types of measurement primitives, such as the raw image data and filtered image data, including their derivatives, edges, lines, corners, and such. As the neighboring receptive fields are forced to have the same directional preference, we assume that neurons with similar velocity selectivity in the neighboring hypercolumns tend to affect each other through receiving inputs from each other as shown in Figure 6.7. This feature implies the smoothness constraint which can be seen more clearly if the network is organized in a multi-layer fashion. Figure 6.8 shows a multi-layer network which is equivalent to the original one. The network consists of $(2D_k+1) \times (2D_l+1)$ layers. Each layer corresponds to a different velocity and contains $N_r \times N_c$ neurons. Each neuron receives excitatory and inhibitory inputs from itself and other neurons in a neighborhood in the same layer. For each point, only the neuron that has the maximum excitation among all neurons in the other layers is on while the others are off. When the neuron at the point (i, j) in the kth and lth layers is in active

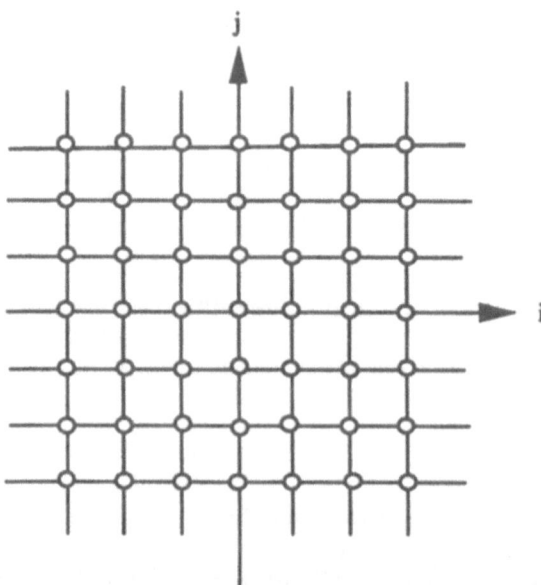

FIGURE 6.6. An alternative grouping pattern for the neurons within a hyper-column. The neurons are uniformly distributed on a two-dimensional Cartesian coordinate plane. Each circle denotes one neuron.

state, this means that the velocities in k and l directions at the point (i,j) are $k\,W$ and $l\,W$, respectively.

Formally, the multi-layer network can be described as follows. Let

$$\{v_{i,j,k,l}, 1 \leq i \leq N_r, 1 \leq j \leq N_c, -D_k \leq k \leq D_k, -D_l \leq l \leq D_l\}$$

be a binary state set V of the neural network with $v_{i,j,k,l}$ denoting the state of the (i,j,k,l)th neuron which is located at point (i,j) in the (k,l)th layer, let $T_{i,j,k,l;m,n,k,l}$ be the synaptic interconnection strength from neuron (i,j,k,l) to neuron (m,n,k,l), and $I_{i,j,k,l}$ be the bias input.

At each step, each neuron (i,j,k,l) synchronously receives inputs from neighboring neurons including itself, and a bias input from the outside world

$$u_{i,j,k,l} = \sum_{(m-i,n-j)\in S_0} T_{i,j,k,l;m,n,k,l} v_{m,n,k,l} + I_{i\,j,k,l} \qquad (6.30)$$

where S_0 is an index set for all neighbors in a $\Gamma \times \Gamma$ window centered at point (i,j). The potential of the neuron, $u_{i,j,k,l}$, is then fed back to corresponding neurons after maximum evolution. The neuron evaluation will be terminated if the network converges.

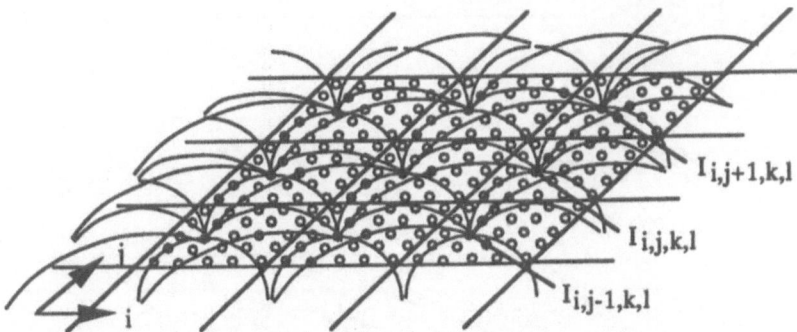

FIGURE 6.7. An artificial neural network. The small frame denotes the hyper-column. The neurons in a hypercolumn are uniformly distributed on a plane. Each neuron receives inputs from itself and other similar directionally selective neurons at the neighboring points, and a bias input from the outside world.

6.3.3   COMPUTING FLOW FIELD

Without adding any physical constraints to the solution, optical flow computed from a pair of image frames usually is noisy and inaccurate. For instance, although the correlation method may provide a solution based on the local match without any smoothness constraint, the resulting optical flow is not accurate and local error is undetectable. In our algorithm, a smoothness constraint is used for obtaining a smooth optical flow field and a line process is employed for detecting motion discontinuities. The line process consists of vertical and horizontal lines, L^v and L^h. Each line can be in either one of two states: 1 for acting and 0 for resting. The error function for computing the optical flow can be properly expressed as

$$
E = \sum_{i=1}^{N_r}\sum_{j=1}^{N_c}\sum_{k=-D_i}^{D_i}\sum_{l=-D_j}^{D_j} \{A\,[(k_{11}(i,j) - k_{21}(i+k, j+l))^2 + (k_{12}(i,j)
$$

$$
- k_{22}(i+k, j+l))^2] + (g_1(i,j) - g_2(i+k, j+l))^2\}v_{i,j,k,l}
$$

$$
+ \frac{B}{2}\sum_{i=1}^{N_r}\sum_{j=1}^{N_c}\sum_{k=-D_i}^{D_i}\sum_{l=-D_j}^{D_j}\sum_{s\in S}(v_{i,j,k,l} - v_{(i,j)+s,k,l})^2
$$

$$
+ \sum_{i=1}^{N_r}\sum_{j=1}^{N_c}\sum_{k=-D_i}^{D_i}\sum_{l=-D_j}^{D_j}\{\frac{C}{2}[(v_{i,j,k,l} - v_{i+1,j,k,l})^2(1 - L^h_{i,j,k,l})
$$

$$
+ (v_{i,j,k,l} - v_{i,j+1,k,l})^2(1 - L^v_{i,j,k,l})] + D(L^h_{i,j,k,l} + L^v_{i,j,k,l})\} \quad (6.31)
$$

where $k_{11}(i,j)$ and $k_{12}(i+k, j+l)$ are the principal curvatures of the first image, $k_{2\bar{1}}(i,j)$ and $k_{22}(i+k, j+l)$ are the principal curvatures of the

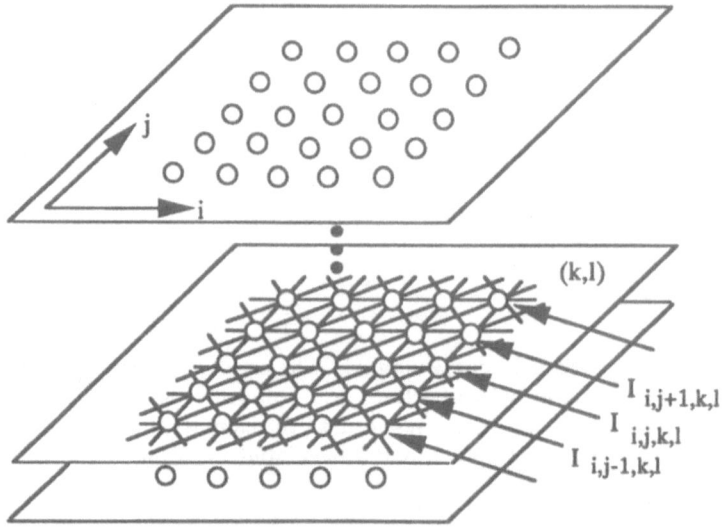

FIGURE 6.8. A multi-layer network which is equivalent to the original one. The neurons are arranged in layers according to their velocity selectivity. i and j denote the image coordinates. k and l denote the velocity coordinates.

second image, $\{g_1(i,j)\}$ and $\{g_2(i+k,j+l)\}$ are the intensity values of the first and second images, respectively, S is an index set excluding $(0,0)$ for all neighbors in a $\Gamma \times \Gamma$ window centered at point (i,j), and A, B, C and D are constants. The first term in (6.31) is to seek velocity values such that all points of two images are matched as closely as possible in a least squares sense. The second term is the smoothness constraint on the solution and the third term is a line process to weaken the smoothness constraint and to detect motion discontinuities. The constants B, C, and D determine the relative importance of the three terms and the constant A in the first term determines the relative importance of the intensity values and their principal curvatures to achieve the best possible results.

By choosing the interconnection strengths and bias inputs as

$$
\begin{aligned}
T_{i,j,k,l;m,n,p,q} = \; & -[48B + C(4 - L^h_{i,j,k,l} - L^h_{i,j+(-1),k,l} - L^v_{i,j,k,l} \\
& -L^v_{i+(-1),j,k,l})]\delta_{i,m}\delta_{j,n}\delta_{k,p}\delta_{l,q} \\
& +C[(1 - L^h_{i,j,k,l})\delta_{i,m}\delta_{j+1,n} + (1 - L^v_{i,j,k,l})\delta_{i+1,m}\delta_{j,n} \\
& +(1 - L^h_{i,j+(-1),k,l})\delta_{i,m}\delta_{j+(-1),n} \\
& +(1 - L^v_{i+(-1),j,k,l})\delta_{i+(-1),m}\delta_{j,n}]\delta_{k,p}\delta_{l,q} \\
& +2B \sum_{s \in S} \delta_{(i,j),(m,n)+s}\delta_{k,p}\delta_{l,q} \qquad (6.32)
\end{aligned}
$$

and

$$I_{i,j,k,l} = -A[(k_{11}(i,j) - k_{21}(i+k,j+l))^2 \qquad (6.33)$$
$$+(k_{12}(i,j) - k_{22}(i+k,j+l))^2] - (g_1(i,j) - g_2(i+k,j+l))^2$$

where $\delta_{a,b}$ is the Dirac delta function, and ignoring the term $D(L^h_{i,j,k,l} + L^v_{i,j,k,l})$ (which does not contain neurons $v_{i,j,k,l}$), the error function (6.31) is transformed into

$$E = -\sum_{i=1}^{N_r}\sum_{j=1}^{N_c}\sum_{k=-D_i}^{D_i}\sum_{l=-D_j}^{D_j}(\frac{1}{2}\sum_{m=1}^{N_r}\sum_{n=1}^{N_c}\sum_{p=-D_i}^{D_i}\sum_{q=-D_j}^{D_j}T_{i,j,k,l;m,n,p,q}v_{m,n,p,q}$$
$$+I_{i,j,k,l})v_{i,j,k,l}, \qquad (6.34)$$

which is the same as (2.7), the energy function of the neural network. Therefore, computation of optical flow can be carried out by neuron evaluation. The size of the smoothing window used in (6.32) is 5.

However, the energy function (6.34) does not include the term $D(L^h_{i,j,k,l} + L^v_{i,j,k,l})$. If only (6.30) is used to update the neurons, the lines cannot be updated properly. It is necessary to update the neurons and lines separately.

Each neuron synchronously evaluates its state and readjusts according to (6.30). The synchronous updating scheme can be implemented in parallel.

The initial state of the neurons is set as

$$v_{i,j,k,l} = \begin{cases} 1 & if \ I_{i,j,k,l} = \max_{p,q}(I_{i,j,p,q}) \\ 0 & otherwise \end{cases} \qquad (6.35)$$

where $I_{i,j,k,l}$ is the bias input. If there are two maximal bias inputs at point (i,j), then only the neuron corresponding to the smallest velocity is initially set at 1 and the other one is set at 0. This is consistent with the Minimal Mapping Theory [Ull79]. In the updating scheme, we also use the minimal mapping theory to handle the case of two neurons having the same largest inputs.

Again, self–feedback may increase the energy function E after a transition. For two neurons $v_{i,j,k,l}$ and $v_{i,j,k',l'}$ changing their states, let the state changes be denoted as

$$\Delta v_{i,j,k,l} = v^{new}_{i,j,k,l} - v^{old}_{i,j,k,l}$$

$$\Delta v_{i,j,k',l'} = v^{new}_{i,j,k',l'} - v^{old}_{i,j,k',l'}$$

and accordingly the energy change ΔE is

$$\Delta E = E^{new} - E^{old} \qquad (6.36)$$

which can be written as

$$
\Delta E = -\sum_{m=1}^{N_r} \sum_{n=1}^{N_c} \sum_{p=-D_i}^{D_i} \sum_{q=-D_j}^{D_j} (T_{i,j,k,l;m,n,p,q}\, \Delta v_{i,j,k,l}
$$

$$
+ T_{i,j,k',l';m,n,p,q}\, \Delta v_{i,j,k',l'}) - I_{i,j,k,l}\, \Delta v_{i,j,k,l} - I_{i,j,k',l'}\, \Delta v_{i,j,k',l'}
$$

$$
- \frac{1}{2} T_{i,j,k,l;i,j,k,l}\, (\Delta v_{i,j,k,l})^2 - \frac{1}{2} T_{i,j,k'.l';i,j,k',l'}\, (\Delta v_{i,j,k',l'})^2
$$

$$
- T_{i,j,k,l;i,j,k',l'}\, (\Delta v_{i,j,k,l} v_{i,j,k',l'}^{new} + \Delta v_{i,j,k',l'} v_{i,j,k,l}^{new}). \tag{6.37}
$$

When

$$
v_{i,j,k,l}^{old} = 0, \qquad v_{i,j,k',l'}^{old} = 1
$$

and

$$
u_{i,j,k',l'} < u_{i,j,k,l},
$$

we have

$$
v_{i,j,k,l}^{new} = 1, \qquad v_{i,j,k',l'}^{new} = 0,
$$

and

$$
\Delta E = (u_{i,j,k',l'} - u_{i,j,k,l}) - \frac{1}{2}\,(T_{i,j,k,l;i,j,k,l} + T_{i,j,k',l';i,j,k',l'}). \tag{6.38}
$$

Since in (6.38), the first term $(u_{i,j,k',l'} - u_{i,j,k,l})$ is negative and the second term $-\frac{1}{2}(T_{i,j,k,l;i,j,k,l} + T_{i,j,k',l';i,j,k',l'})$ is

$$
48\, B + C\, (4 - L_{i,j,k,l}^{h} - L_{i,j+(-1),k,l}^{h} - L_{i,j,k,l}^{v} - L_{i+(-1),j,k,l}^{v}) > 0,
$$

if the first term is less than the second term, then $\Delta E > 0$. A deterministic decision rule is used for updating the neuron states.

For updating the lines, the following decision rule is used. Let $L_{i,j,k,l}^{;new}$ and $L_{i,j,k,l}^{;old}$ denote the new and old states of the line $L_{i,j,k,l}$, respectively. By (6.31), the energy changes due to the state changes of the vertical line $L_{i,j,k,l}^{v}$ and horizontal line $L_{i,j,k,l}^{h}$ can be determined:

$$
\begin{aligned}
\Delta E^v &= E^{new} - E^{old} \\
&= \sum_{i=1}^{N_r} \sum_{j=1}^{N_c} \sum_{k=-D_i}^{D_i} \sum_{l=-D_j}^{D_j} \left[\frac{C}{2}(v_{i,j,k,l} - v_{i+1,j,k,l})^2 (L_{i,j,k,l}^{v;old} - L_{i,j,k,l}^{v;new}) \right. \\
&\quad \left. + D(L_{i,j,k,l}^{v;new} - L_{i,j,k,l}^{v;old}) \right]
\end{aligned} \tag{6.39}
$$

and

$$
\begin{aligned}
\Delta E^h &= E^{new} - E^{old} \\
&= \sum_{i=1}^{N_r} \sum_{j=1}^{N_c} \sum_{k=-D_i}^{D_i} \sum_{l=-D_j}^{D_j} \left[\frac{C}{2}(v_{i,j,k,l} - v_{i,j+1,k,l})^2 (L_{i,j,k,l}^{h;old} - L_{i,j,k,l}^{h;new}) \right. \\
&\quad \left. + D(L_{i,j,k,l}^{h;new} - L_{i,j,k,l}^{h;old}) \right],
\end{aligned} \tag{6.40}
$$

respectively. Then, the vertical line $L^v_{i,j,k,l}$ and the horizontal line $L^h_{i,j,k,l}$ take a new state if the energy changes ΔE^v and ΔE^h are less than zero, respectively.

The algorithm for computing the optical flow can then be summarized as follows:

1. Set the initial state of the neurons.

2. Update the state of all the lines synchronously.

3. Update the state of all neurons synchronously.

4. Check the energy function. If the energy does not change anymore, stop; otherwise, go back to Step 2.

6.4 Detection of Motion Discontinuities

Motion discontinuities in optical flow often result from occluding contours of moving objects. In this case, the moving objects are projected into the image plane as adjacent surfaces and their boundaries undergo either split, fusion, or non-split-fusion motion. These motion situations give rise to discontinuities along their boundaries. We must detect these discontinuities to prevent the algorithm from using the physical constraint of surface smoothness in computating the optical flow from one surface to the other. Basically there are two approaches for detecting motion discontinuities. The first approach locates significant changes of optical flow from the computed optical flow field [Pot75, Tho80, TMB85, Hil83]. In [TMB85], a zero-crossing edge detector is used to find the discontinuities in a dense optical flow generated from a sparse one by interpolation. This scheme allows surfaces to translate and rotate as well. Instead of directly detecting discontinuities, region growing techniques are employed in [Pot75, Tho80] to group elements of similar velocities. The discontinuities are then implicitly given by the boundaries between the regions. Hildreth [Hil83] utilizes the properties of the initial perpendicular components of velocity to locate the discontinuities along zero-crossing contours derived from the image. These schemes are limited to pure translational motion. The second approach is to infer discontinuities from the initial motion measurements without fully computing the optical flow field [Hor86, Koc87]. In fact this approach interleaves detection of discontinuities and computation of optical flow. By incorporating a line process into the optical flow equation, Koch [Koc87] gives an explicit formulation for detecting the discontinuities. Although these methods can infer discontinuities, the location of discontinuities due to fusion motion may be shifted. In order to locate the discontinuities more accurately, we design a method for detecting the occluding elements based on the initial

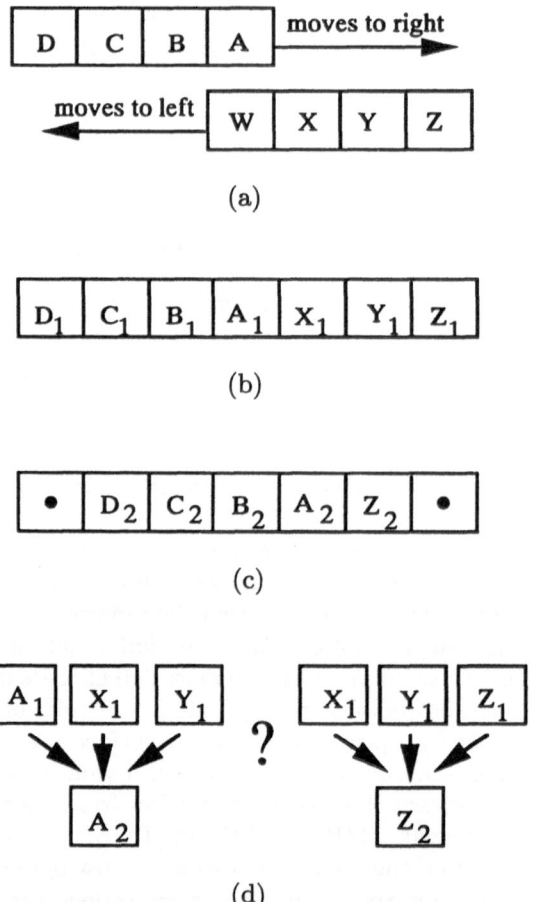

FIGURE 6.9. Fusion motion. (a) Two objects move towards each other. (b) First frame. (c) Second frame. (d) Fusion competition.

motion measurements. Once the information about the occluding elements is available, the neural network will correctly locate motion discontinuities.

To formalize the analysis, we must distinguish split motion, fusion motion and non-split-fusion motion. As explained in [Ull79], the split motion occurs when a single element is replaced by multiple elements, as in two successive frames where the single element is shown first followed by multiple elements, while the fusion motion results when multiple elements are presented first followed by a single element. The non-split-fusion motion does not give new elements or eliminate old elements; the number of elements between the two frames does not change. Two simple examples are given in Figures 6.9 and 6.10 to illustrate split motion and fusion motion, respectively. Each example is composed of two frames and each frame contains two moving surfaces,

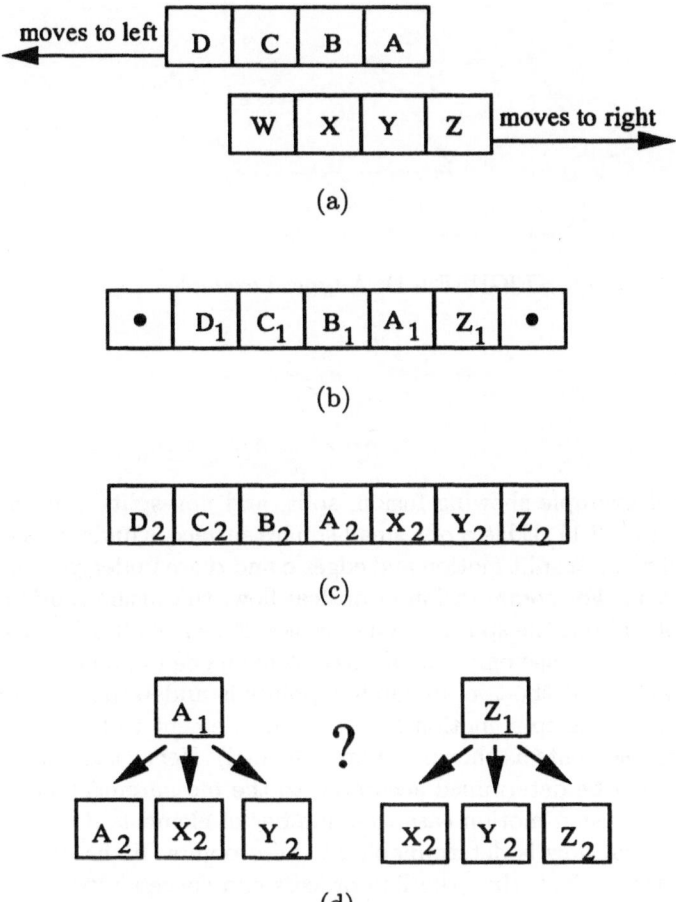

FIGURE 6.10. Split motion. (a) Two objects move away from each another. (b) First frame. (c) Second frame. (d) Split competition.

one in front and the another in back. For the first frame, the elements of the surface in the front are denoted by A_1, B_1, and so forth, and the elements of the surface in back by W_1, X_1, and so forth. For the second frame, A_2, B_2, and so forth and W_2, X_2, and so forth denote the elements of the surface in front and the surface in back, respectively. The elements can be either image pixels or lines. In Figure 6.9 two surfaces are moving towards each other. The elements X and Y of the surface in back are visible in the first frame, while in the second frame they are occluded by the front surface. Hence, the elements A_1, X_1 and Y_1 (or X_1, Y_1 and Z_1) are replaced by the element A_2 (or Z_2), and a fusion motion is observed. In Figure 6.10 two surfaces are moving away from one another. The element A_1 (or Z_1) is presented by A_2, X_2 and Y_2 (or X_2, Y_2 and Z_2), and a split motion is observed.

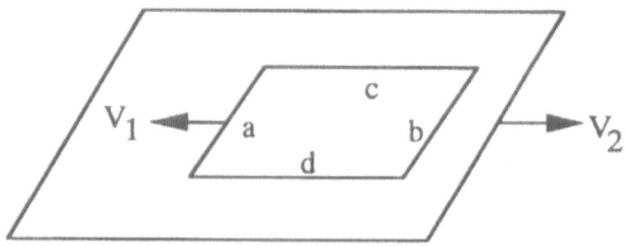

FIGURE 6.11. A typical example.

One typical example showing fusion, split, and non-split-fusion motion is given in Figure 6.11. Edge **a** of a small square surface is undergoing a fusion motion, edge **b** is a split motion and edges **c** and **d** are undergoing non-split-fusion motion. For computation of optical flow, the fusion motion usually causes problem but the split motion does not. As shown in Figure 6.9(d), in the fusion motion case only one of three elements can find correspondence. The optical flow at the two unmatched points is undetermined. Unlike the fusion motion, the split motion has only one element to be matched with one of three elements as shown in Figure 6.10(d). Hence the optical flow at that point can be determined according to the measurement primitive. In the non-split-fusion motion case, the number of elements does not change and the optical flow is determinable at these points. If the optical flow is perfectly determined, then the line process can successfully and correctly locate motion discontinuities. Thus we will concentrate on the fusion motion case for detecting the occluding elements.

Suppose that the surfaces are translating with constant velocities. Let us consider the case in which a surface is moving against a stationary background as shown in Figure 6.12. Let X_1 denote the occluding element, A_2 and X_2 the corresponding elements of A_1 and X_1, respectively. Let (i, j) be the coordinates of element A_1, and d_i and d_j the i and j components of optical flow at (i, j), respectively. We assume that X_1 and Y_1 are located at $(i + d_i, j + d_j)$ and $(i + 2 \times d_i, j + 2 \times d_j)$, respectively. By defining the matching errors $\mathbf{e}_1(i, j)$, $\mathbf{e}_2(i, j)$, $\mathbf{e}_3(i, j)$, and $\mathbf{e}_4(i, j)$ as

$$\mathbf{e}_1(i, j) = I_{i,j,d_i,d_j}$$

$$\mathbf{e}_2(i, j) = I_{i+d_i,j+d_j,0,0}$$

$$\mathbf{e}_3(i, j) = I_{i+d_i,j+d_j,d_i,d_j}$$

$$\mathbf{e}_4(i, j) = I_{i+2\times d_i,j+2\times d_j,0,0}$$

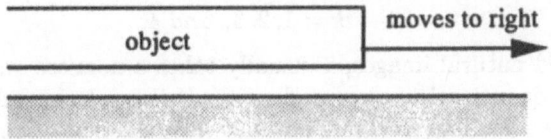

Stationary background

(a)

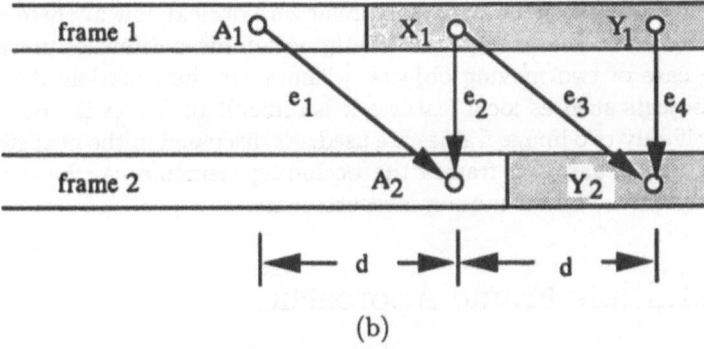

(b)

FIGURE 6.12. Detection of the occluding element. (a) One object moves against a stationary background. (b) Detection scheme.

where $I_{.,.,.,.}$ are bias inputs given in (6.33), the following relations hold under orthographic or perspective projection without motion along the optical axis:

$$\begin{aligned} \mathbf{e}_1(i,j) &\leq \mathbf{e}_2(i,j) \\ \mathbf{e}_4(i,j) &\leq \mathbf{e}_3(i,j). \end{aligned} \tag{6.41}$$

Note that if the above relations do not hold then the element X_1 is not an occluding element. Hence, it is natural to use the relations (6.41) for detecting the occluding elements.

Detection rule: An occluding element is detected at $(i + d_i, j + d_j)$ if the optical flow has nonzero values at (i, j) and

$$\begin{aligned} \bar{\mathbf{e}}_2(i.j) - \bar{\mathbf{e}}_1(i,j) &> T \\ \bar{\mathbf{e}}_4(i.j) - \bar{\mathbf{e}}_3(i,j) &> T \end{aligned} \tag{6.42}$$

where the theshold T is a nonnegative number and $\bar{\mathbf{e}}_k(i, j)$ are the average values of the matching errors within a $\Gamma_T \times \Gamma_T$ window S_T

$$\bar{\mathbf{e}}_k(i, j) = \frac{1}{\Gamma_T^2} \sum_{s \in S_T} \bar{\mathbf{e}}_k((i, j) + s)$$

$$\text{for} \quad k = 1, 2, 3, \text{ and } 4.$$

For digitized natural images, T usually takes a nonzero value to reduce the effects of quantization error and noise. Using a large value for T can eliminate false occluding elements but may miss the true ones. Since optical flow at an occluding point has zero values, the *a priori* knowledge about the occluding elements can be embedded in the bias inputs by setting (for instance at point $(i + d_i, j + d_j)$)

$$I_{i+d_i,j+d_j,0,0} = min(I_{i+d_i,j+d_j,k,l}; \ -D_i \leq k \leq D_i, -D_j \leq l \leq D_j). \quad (6.43)$$

Accordingly, the neural network will prefer zero optical flow at these points and therefore the line process can locate motion discontinuities precisely.

In the case of two moving objects, without any information about the moving objects such as local features it is difficult to detect the occluding elements if only two image frames are used. As discussed in the next section, by using more than two frames the occluding elements can be detected based only on the initial motion measurements.

6.5 Multiple Frame Approaches

When the image quality is poor, measurement primitives estimated from these images are not accurate and reliable. For instance, if the images are blurred by motion, then the local features are smeared and much of the information is lost. Especially, the object boundaries become wide and the derivatives of the intensity function become small at the boundaries. Based on these low quality measurements, motion discontinuities cannot be correctly located and hence optical flow cannot be accurately computed. To improve the accuracy, one way is to improve the image quality by using some image restoration techniques to remove degradations. However, without *a priori* knowledge of the degradations, such as blur function, an image cannot be restored perfectly. When the blur is ill conditioned, it is still difficult to restore the image even if the blur function is known. An alternative is to compute optical flow over a long time interval, by using multiple frames. In this section, two algorithms, batch and recursive, using more than two frames of images are presented.

6.5.1 BATCH APPROACH

We assume that the object in the scene is under a translational motion and M frames of images are available. The motion velocity is also assumed to be a constant. It is interesting to note that in the two frame case the bias inputs (6.33) contain nothing but the measurement primitives, the intensity values, and principal curvatures, which are estimated from images. The bias inputs of the network are completely determined by the observations, or the images, while the interconnection strengths (6.32) do not contain any observations. All these facts suggest that any information from the outside world can be included in the bias inputs. The network learns all information directly from the inputs. Hence, it is natural to extend the two frame approach to multiple frames by adding more observations to the bias inputs:

$$
\begin{aligned}
I_{i,j,k,l} = \ & -\sum_{r=1}^{M-1} \{ A[(k_{r1}(i+(r-1)k, j+(r-1)l) \\
& -k_{(r+1)1}(i+rk, j+rl))^2 + (k_{r2}(i+(r-1)k, j+(r-1)l) \\
& -k_{(r+1)2}(i+rk, j+rl))^2] + (g_r(i+(r-1)k, j+(r-1)l) \\
& -g_{r+1}(i+rk, j+rl))^2 \}.
\end{aligned}
\tag{6.44}
$$

Accordingly, the initial state of the neurons (6.35) is set by using these new bias inputs. The two frame algorithm can then be used without any modifications for the multiple frame case.

6.5.2 RECURSIVE ALGORITHM

If all the images are not available at the same time or if one wants to compute the optical flow in real time, a recursive algorithm can be used. The recursive algorithm uses an RLS algorithm. to update the bias inputs. First, the initial condition for the bias inputs is set to zero:

$$
I_{i,j,k,l}(0) = 0.
\tag{6.45}
$$

This is reasonable, because there is no information available at the beginning. Then, whenever a new frame becomes available, the bias inputs can be updated by

$$
I_{i,j,k,l}(r) = I_{i,j,k,l}(r-1) + (1/r)(\tilde{I}_{i,j,k,l}(r) - I_{i,j,k,l}(r-1))
\tag{6.46}
$$

$$
\text{for} \quad 2 \le r \le M
$$

where $\tilde{I}_{i,j,k,l}(r)$ is a new observation given by

$$
\begin{aligned}
\tilde{I}_{i,j,k,l}(r) = \ & -A[(k_{r1}(i+(r-1)k, j+(r-1)l) \\
& -k_{(r+1)1}(i+rk, j+rl))^2 + (k_{r2}(i+(r-1)k, j+(r-1)l) \\
& -k_{(r+1)2}(i+rk, j+rl))^2] + (g_r(i+(r-1)k, j+(r-1)l) \\
& -g_{r+1}(i+rk, j+rl))^2.
\end{aligned}
\tag{6.47}
$$

In fact, this RLS algorithm is equivalent to the batch algorithm. If the intermediate results are not required, the optical flow can be computed after all images are received. As one can see, the RLS algorithm is parallel in nature and very few computations are required at each step. Hence this algorithm is extremely fast and can be implemented in real time.

6.5.3 DETECTION RULES

Since the number of occluding elements dramatically increases as more image frames are involved, special attention has to be paid to this problem. With minor modifications, the detection criterion used for the two frames can be extended to multiple frames. Define the matching errors as

$$
\begin{aligned}
\mathbf{e}_{1,r}(i,j) = {} & A[(k_{11}(i,j) - k_{(r+1)1}(i+r_{0i},j+r_{0j}))^2 \\
& + (k_{12}(i,j) - k_{(r+1)2}(i+r_{0i},j+r_{0j}))^2] \\
& + (g_1(i,j) - g_{r+1}(i+r_{0i},j+r_{0j}))^2 \qquad (6.48)
\end{aligned}
$$

$$
\begin{aligned}
\mathbf{e}_{2,r}(i,j) = {} & A[(k_{11}(i+r_{0i},j+r_{0j}) - k_{(r+1)1}(i+r_{0i},j+r_{0j}))^2 \\
& + (k_{12}(i+r_{0i},j+r_{0j}) - k_{(r+1)2}(i+r_{0i},j+r_{0j}))^2] \\
& + (g_1(i+r_{0i},j+r_{0j}) - g_{r+1}(i+r_{0i},j+r_{0j}))^2 \quad (6.49)
\end{aligned}
$$

$$
\begin{aligned}
\mathbf{e}_{3,r}(i,j) = {} & A[(k_{11}(i+r_{0i},j+r_{0j}) - k_{(r+1)1}(i+r_{1i},j+r_{1j}))^2 \\
& + (k_{12}(i+r_{0i},j+r_{0j}) - k_{(r+1)2}(i+r_{1i},j+r_{1j}))^2] \\
& + (g_1(i+r_{0i},j+r_{0j}) - g_{r+1}(i+r_{1i},j+r_{1j}))^2 \quad (6.50)
\end{aligned}
$$

$$
\begin{aligned}
\mathbf{e}_{4,r}(i,j) = {} & A[(k_{11}(i+r_{1i},j+r_{1j}) - k_{(r+1)1}(i+r_{1i},j+r_{1j}))^2 \\
& + (k_{12}(i+r_{1i},j+r_{1j}) - k_{(r+1)2}(i+r_{1i},j+r_{1j}))^2] \\
& + (g_1(i+r_{1i},j+r_{1j}) - g_{r+1}(i+r_{1i},j+r_{1j}))^2 \quad (6.51)
\end{aligned}
$$

where

$$
r_{0i} = rd_i,
$$

$$
r_{0j} = rd_j,
$$

$$
r_{1i} = (r+1)d_i,
$$

$$
r_{1j} = (r+1)d_j,
$$

and d_i and d_j are the i and j components of optical flow at (i,j). The detection rule in the case of one moving object is similar to (6.42). For the batch algorithm, the detection rule is given by

Detection rule 1: An occluding element is detected at $(i+(r+1)d_i, j+(r+1)d_j)$ if the optical flow at (i,j) is not zero and

$$
\begin{aligned}
\bar{e}_{2,k}(i.j) - \bar{e}_{1,k}(i,j) &> T, \\
\bar{e}_{4,k}(i.j) - \bar{e}_{3,k}(i,j) &> T
\end{aligned} \qquad (6.52)
$$

$$
\text{for} \quad 0 < k \le r \le M - 1
$$

where $\bar{e}_{.,k}(i.j)$ are the avarage values of the matching error.

When only a few images are available, we cannot detect all the occluding elements. Hence, the detection rule for the recursive algorithm is modified as

Detection rule 2: When the $(r+1)$th image frame becomes available, an occluding element will be detected at $(i + (r+1)d_i, j + (r+1)d_j)$ if the optical flow at (i,j) is not zero and

$$\bar{e}_{2,r}(i.j) - \bar{e}_{1,r}(i,j) > T,$$
$$\bar{e}_{4,r}(i.j) - \bar{e}_{3,r}(i,j) > T. \qquad (6.53)$$

Once an occluding element is detected at (i,j), all the bias inputs corresponding to that point will be reset and will not be updated anymore.

In the case of two moving objects as shown in Figure 6.13 (a), occluding elements can be detected by using the multiple frames. In the second frame, region 1 is occluded and in the third frame both regions 1 and 2 are occluded. Hence, all the elements of regions 1 and 2 are occluding elements. The detection scheme is as follows. As shown in Figure 6.13 (b) (for simplicity, we only illustrate the detection scheme for the one-dimensional case), U_1 and V_1 belong to region 1 and W_1 and X_1 to region 2. Using the first two frames, optical flow at A_1 and W_1 can be determined if the matching errors between A_1 and A_2 and W_1 and W_2 are smaller than some threshold. Similarly, optical flow at A_2 and Y_2 can be computed by using the frames 2 and 3. If the optical flow at points W_1, Y_1, W_2 and Y_2 is the same but different from that at point A_1, then W_1 and X_1 are occluding elements and therefore U_1 and V_1 are occluding elements also.

The algorithms based on multiple frames can be summarized as follows:

1. Compute optical flow from the first two frames.

2. Update bias inputs.

3. Detect occluding elements and reset bias inputs accordingly. For the recursive algorithm, go back to Step 2 if the incoming frame is not the last one; otherwise, go to Step 4.

4. Compute optical flow with updated bias inputs.

6.6   Experimental Results

A number of synthetic and natural image sequences were tested. For each point, we use two memories in the range $-D_i$ to D_i and $-D_j$ to D_j to represent velocities in i and j directions, respectively, instead of using $(2D_i+1)$ and $(2D_j+1)$ neurons. Due to local connections of the neurons, the neuron input $U_{i,j,k,l}$ is computed only within a small window.

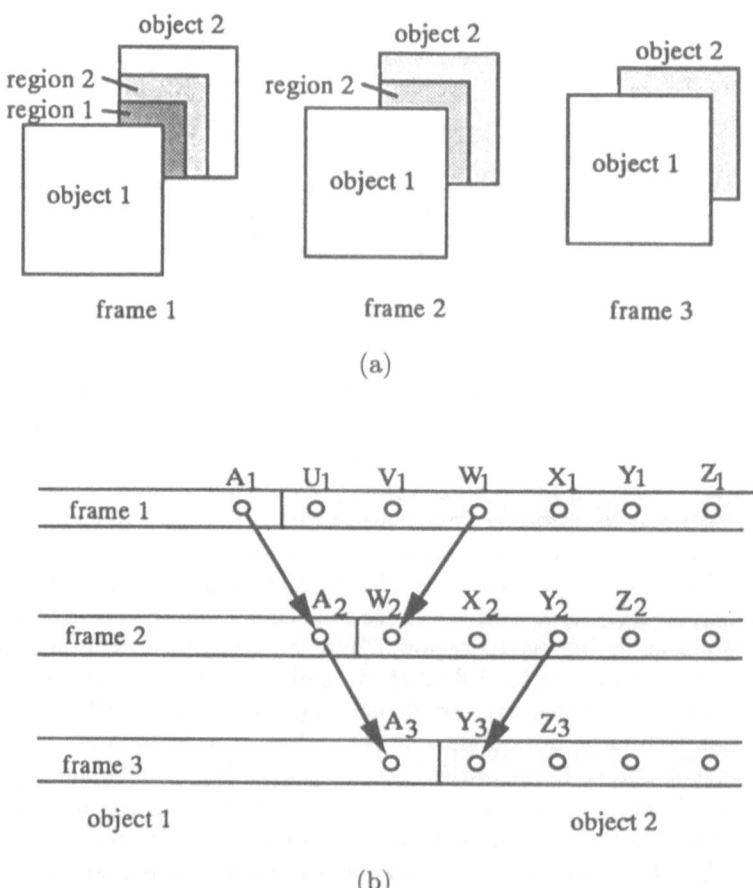

FIGURE 6.13. Detection of occluding elements using multiple frames. (a) Two objects move toward each other. (b) Detection scheme.

6.6.1 SYNTHETIC IMAGE SEQUENCE

Two experiments based on synthetic image sequences, made of purely translational random dot images and rotating disk images, are presented here. The purely translational random dot images were created by the pseudo random number generating method described in [Jul60] used for generating random dot stereograms. Each dot consists of only one element. Figures 6.14(a) and (b) show 10% intensity translational random dot images (only 10% of the pixels are white and 90% are black). Intensity values of the white and black elements were set at 255 and 0, respectively. The images are of size 128×128 and contains two square patches which are moving against a stationary background. A center 20×20 patch is moving in the southeast direction with velocity ($v_i = 3, v_j = 3$), while a center

40×40 patch partially overlapped by the smaller patch is moving in the south-southeast direction with velocity ($v_i = 2, v_j = 1$). Figure 6.14(c) is the resulting optical flow after ten iterations. For display purposes, only a 70×70 center part of the optical flow field is shown in the figure. The velocity values are normalized by the maximum velocity value. We used $A = 20$, $B = 6,000$, $C = 8,000$, $D = 1$, $D_k = D_l = 3$, $\omega = 7$, and $W = 1$. The highest order of the polynomials is four. Note that the flow field is dense.

Another test was run on the rotating disk sequence with intensity values in range (0–255). Figures 6.15(a) and (b) show the first and second frames of intensity images, respectively. Since the rotating disk is not globally rigid and the principal curvatures are estimated over a small window, we assume that the disk is rigid locally. The centered part of optical flow shown in Figure 6.15(c) was obtained after 25 iterations. The parameters used were $A = 4$, $B = 15$, $C = 0$, $D = 1$, $D_k = D_l = 63$, $\omega = 5$, and $W = \frac{1}{9}$. An eighth order polynomial was used to fit data. Since the parameter C was set at zero, no line process was involved.

6.6.2 NATURAL IMAGE SEQUENCE

A sequence of pick-up truck images taken from a static camera was used to test both the nonbatch approach (using two image frames) and the batch approach (using multiple image frames). Figure 6.16 shows four successive frames of the pick-up truck image, a pick-up truck moving from right to left against a stationary background. Since the shutter speed was low, the truck was heavily blurred by the motion. The motion blur smeared the edges and erased local features, especially the features on the wheels. Hence, it is difficult to detect rotation of the wheels. Since the rear part of the truck is missing in the first frame, we reversed the order of the image sequence so that there is a complete truck image in the first frame. Accordingly, the direction of the computed optical flow should be reversed. For the two frame approach, we used the fourth frame as the first frame and the third frame as the second frame. The original images are of size 480×480. For simplicity of computation, the image size was reduced to 120×120 by subsampling. To estimate the principal curvatures and the intensity values, an 11×11 window, thus $\omega = 5$, was chosen and a fourth order polynomial was used for all images. By setting $A = 2$, $B = 250$, $C = 50$, $D = 20$, $D_k = 7$, $D_l = 1$, and $W = 1$, the optical flow was obtained after 36 iterations. A 48×113 sample of the computed optical flow corresponding to the part framed by black lines (in the first frame) in Figure 6.16 is given in Figure 6.17. Note that although most of the boundary locations are correct, the boundaries due to the fusion motion (such as the rear part of the truck and the driver's cab) are shifted by the line process.

Figure 6.18 displays the occluding pixels detected at $T = 100$ based on the initially computed optical flow of Figure 6.17. By embedding the information about the occluding pixels into the bias inputs, using the initially computed optical flow as the initial conditions and choosing $A = 2$, $B = 188$, $C = 200$, $D = 20$, $D_k = 7$, $D_l = 1$, and $W = 1$, the final result shown in Figure 6.19 was obtained after 13 iterations. The accuracy of the boundary location is significantly improved.

For the multiple frame approaches, we used four image frames. Theoretically there is no limit to the number of frames that can be used in the batch approach. For the same reason mentioned before, the fourth frame was taken as the first frame, the third frame as the second frame, etc. As mentioned before, the batch and recursive algorithms are equivalent and a set of identical parameters was used for both algorithms in the experiment, the same results were obtained. Figure 6.20 displays the occluding pixels detected at $T = 100$ from four frames. Figure 6.21 shows the optical flow computed from four frames using the occluding pixel information. The parameters used were $A = 4$, $B = 850$, $C = 80$, $D = 20$, $D_k = 7$, $D_l = 1$, and $W = 1$, and 12 iterations were required. As expected, the output is much cleaner and the boundaries are more accurate than that of the approach based on two frames. The number of iterations is also reduced.

6.7 Discussion

We have presented two neural network-based approaches for computing optical flow. For the approach based on two frames, we made no assumptions or requirements on the solutions except smoothness. For the batch approach, we assumed that the object is undergoing a pure translational motion. Experimental results show that principal curvatures are useful for matching, and our approaches based on such measurement primitives work very well, especially for some low-quality natural images such as the truck images. The algorithm for detecting motion discontinuities also works very well. The experimental results also show that if only part of the occluding elements is detected, then the line process may incorrectly locate the discontinuities based on incomplete information. Figure 6.22 shows the optical flow computed from four frames, using partial information about the occluding elements, thus using Figure 6.17 rather than Figure 6.20.

A common drawback of the neural network-based approachs, or more generally speaking the constrained minimization problems, is the optimal choice of parameters. This is a problem many researchers have been trying to solve for many years, and there is no systematic way for choosing the parameters. However, we found that the smoothing parameter B is crucial and very much depends on the variance of the bias input. When the variance is large, the parameter B should be large too. Hence, we first estimated the variance of the bias input and set B accordingly, and then adjusted it experimentally.

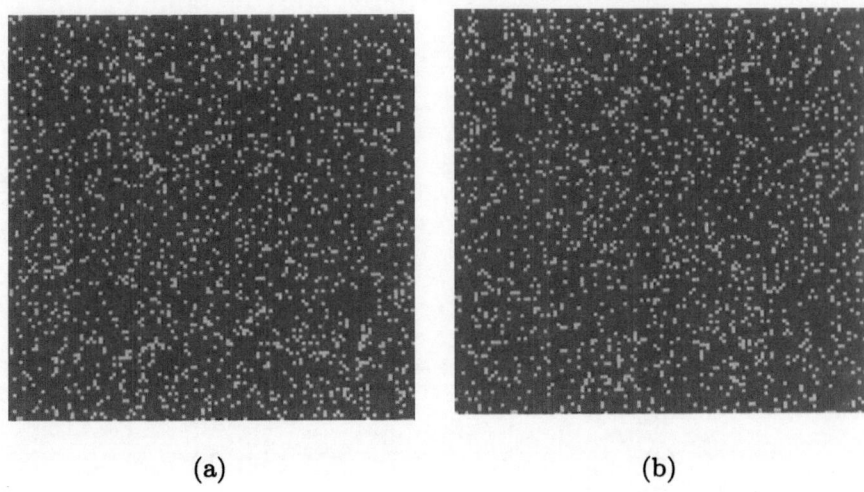

(a)   (b)

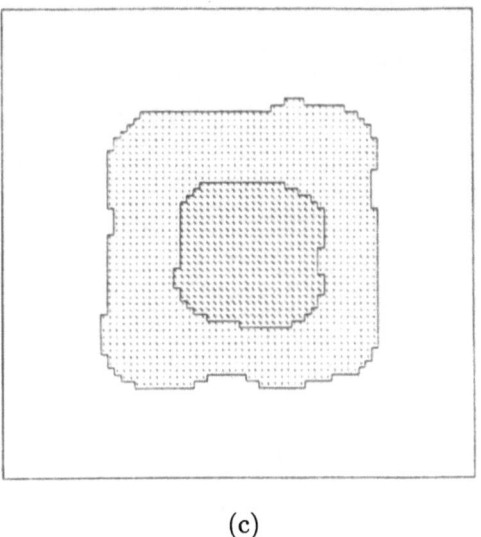

(c)

FIGURE 6.14. 10% density translational random dot images. (a) The first frame. (b) The second frame. (c) Optical flow.

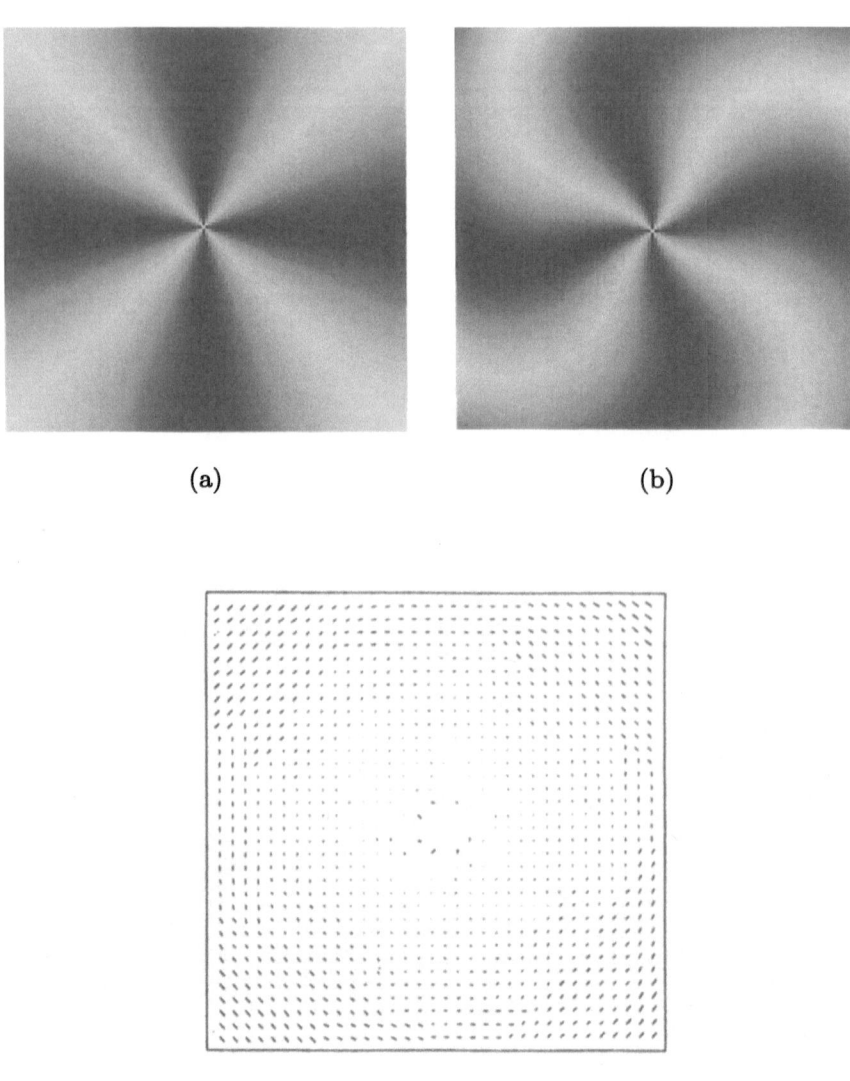

(a)   (b)

(c)

FIGURE 6.15. Nonrigid rotating disk images. (a) The first frame. (b) The second frame. (c) Optical flow.

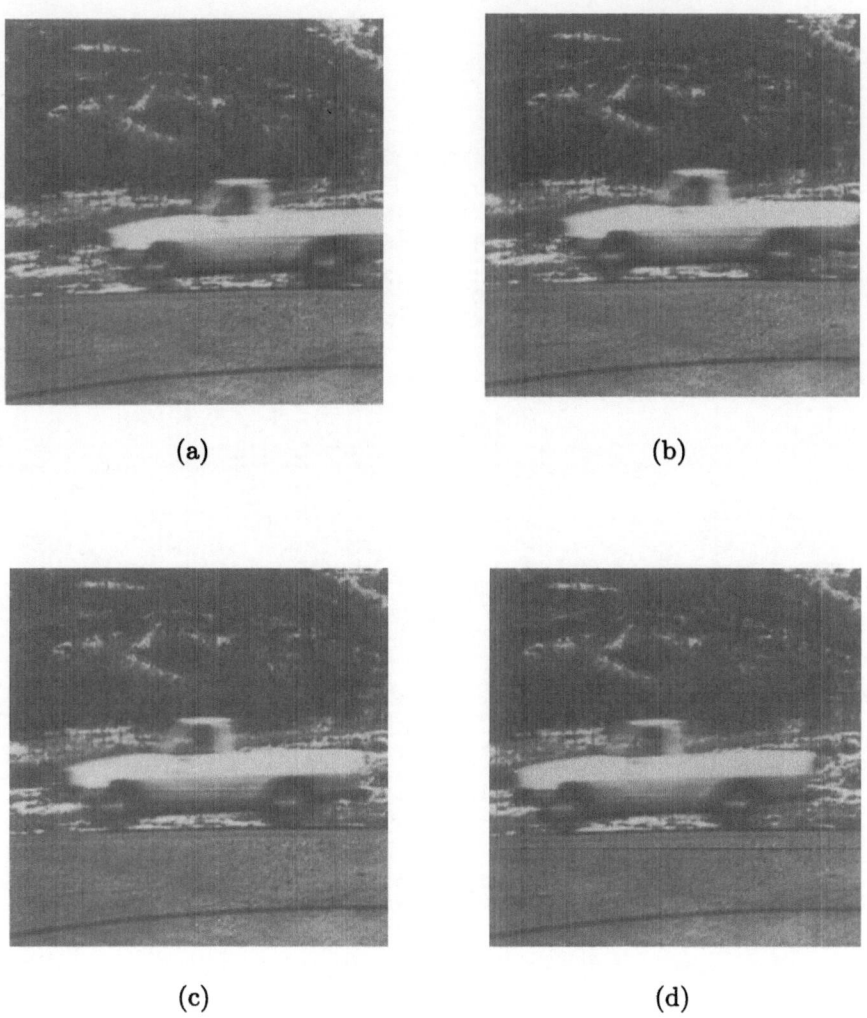

FIGURE 6.16. A sequence of pick-up truck images. (a) The first frame. (b) The second frame. (a) The third frame. (b) The fourth frame.

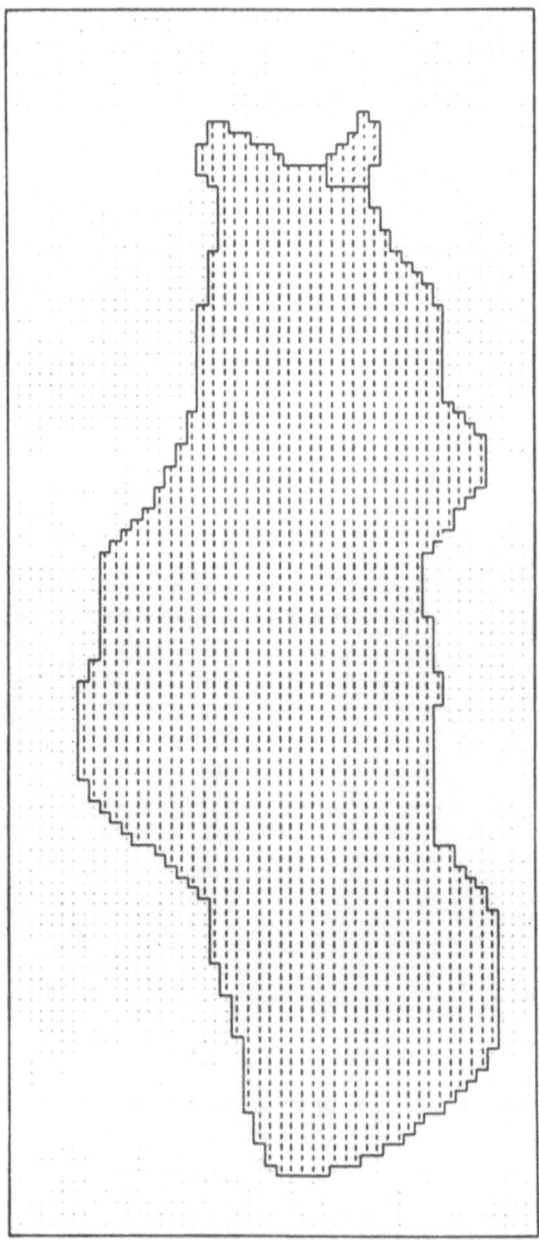

FIGURE 6.17. Optical flow computed from two images. (Reprinted with permission from [ZC90] Figure 8, ©1990 IEEE.)

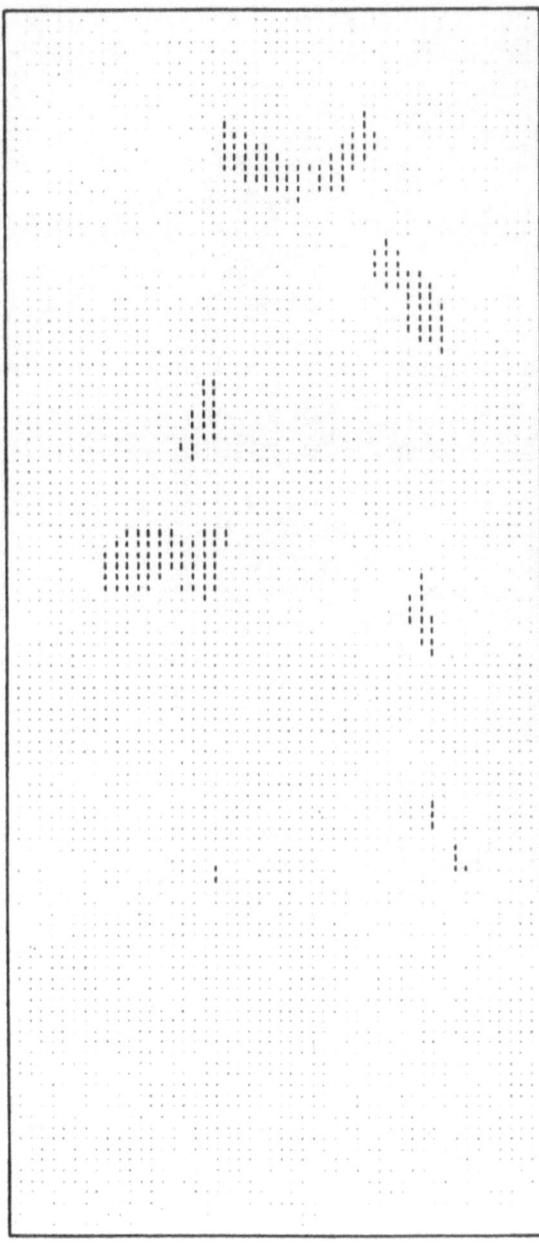

FIGURE 6.18. Occluding pixels detected from two frames.

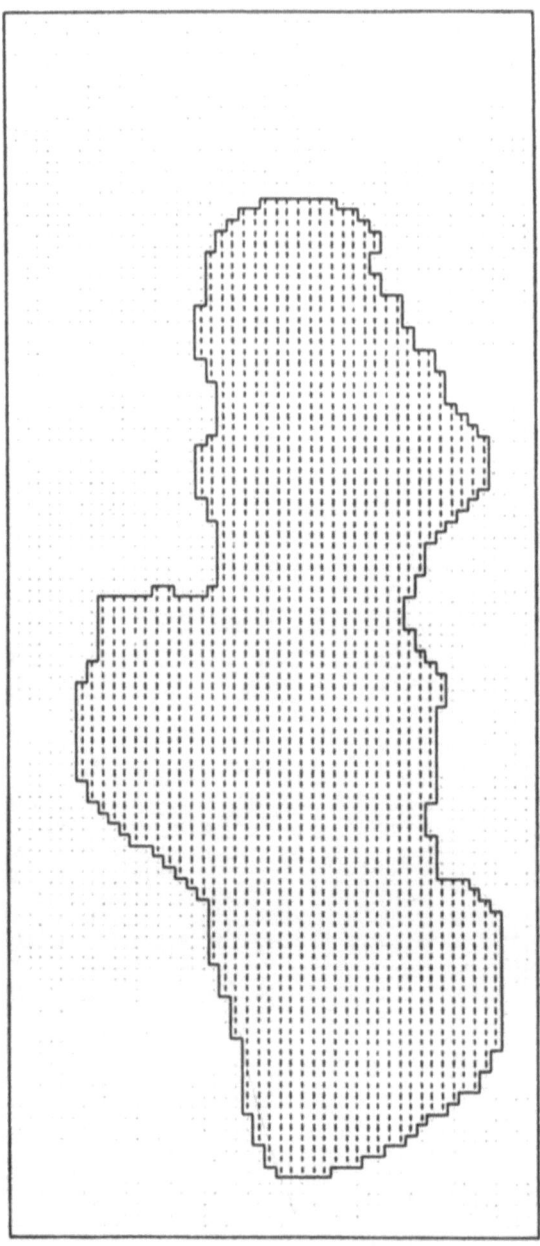

FIGURE 6.19. Optical flow computed from two frames using the information about the occluding pixels. (Reprinted with permission from [ZC90] Figure 9, ©1990 IEEE.)

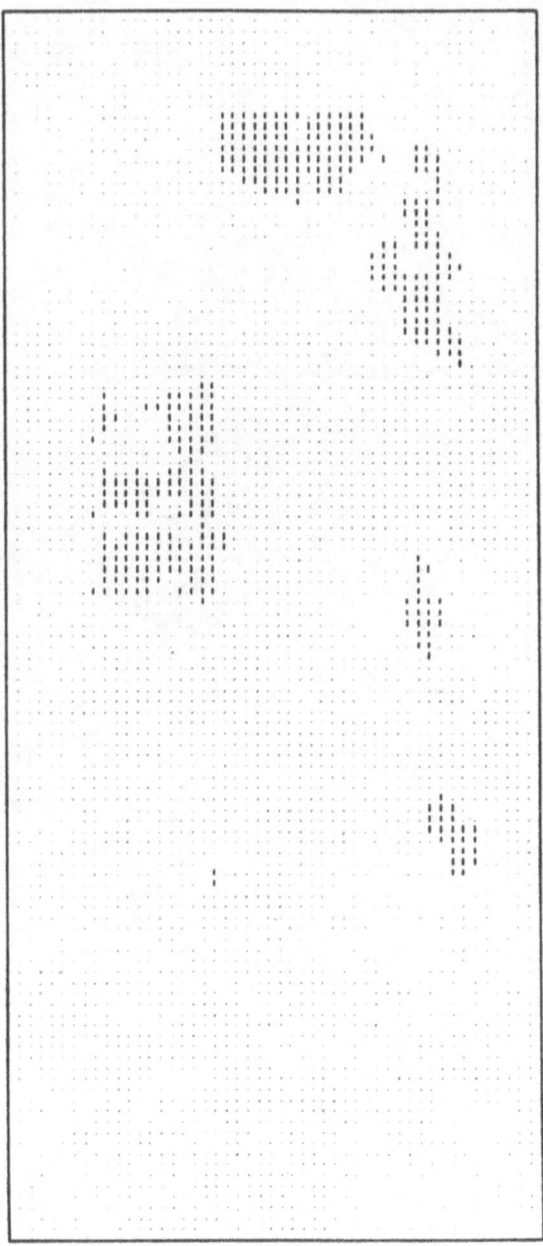

FIGURE 6.20. Occluding pixels detected from four frames.

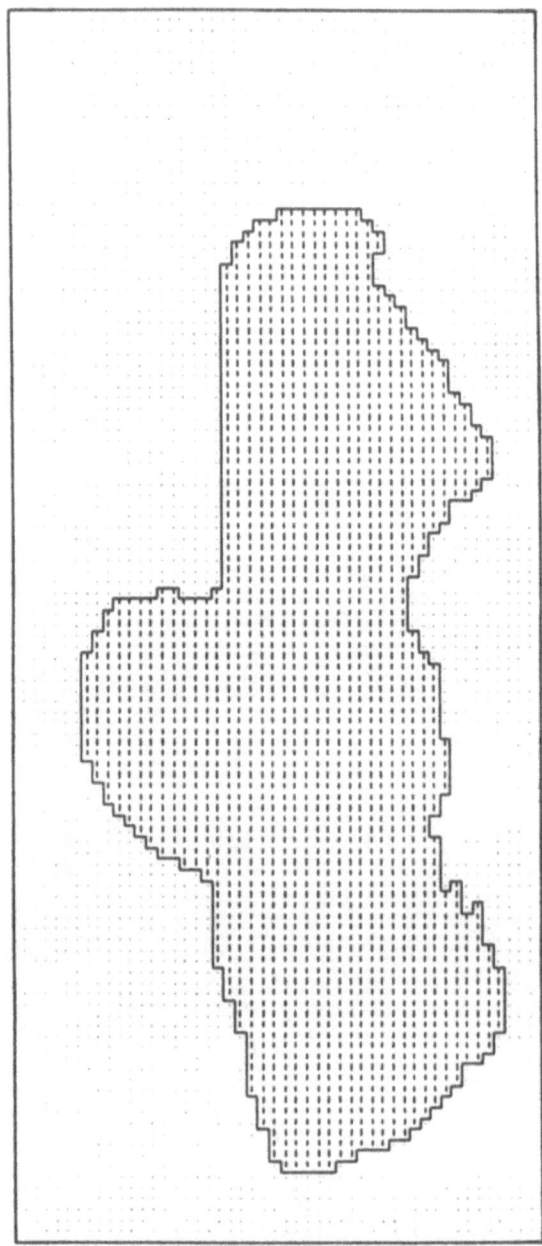

FIGURE 6.21. Optical flow computed from four frames using the information about the occluding pixels. (Reprinted with permission from [ZC90] Figure 10, ©IEEE.)

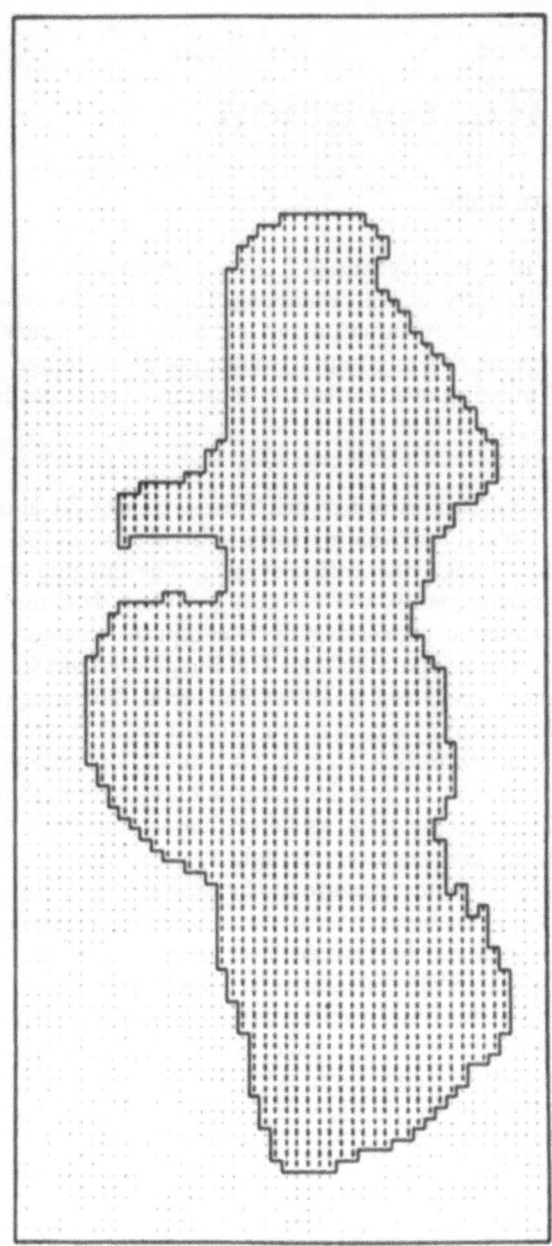

FIGURE 6.22. Optical flow computed from four frames using the partial information about occluding elements.

7

Image Restoration

7.1 Introduction

Restoration of a high quality image from a degraded recording is an important problem in early vision processing. Image usually refers to a two-dimensional light intensity function $x(a, b)$. Since light intensity is a real positive quantity and the maximum brightness of an image is restricted by the practical imaging system, $x(a, b)$ is a finite, real and non-negative function, where

$$0 \leq x(a, b) \leq A$$

with A being the maximum image brightness. Figure 7.1 shows a digital image restoration system containing three subsystems: an imaging system, image digitizer and image restoration system. The imaging system, which consists of an optical system and recording devices, is a major source of degradations. To enable processing by a computer, images are sampled and quantized by the image digitizer. The image digitizer also introduces some degradations because of quantization error. The image restoration system uses some techniques to remove (1) deterministic degradations indexDegradations, deterministic such as blur due to optical system aberrations, diffraction, motion, atmospheric turbulence, film nonlinearities, and (2) statistical degradations such as noise due to electronic imaging sensors, film granularity and atmospheric light fluctuations. The digital image restoration system gives an estimate of the original image in some sense.

Image restoration has been extensively studied. Over the last 20 years, various methods such as the inverse filter [AH77], Wiener filter [AH77], Kalman filter [WI81], SVD pseudoinverse [AH77, Pra78], and many other model-based approaches have been proposed. One of the major drawbacks of most of the image restoration algorithms is the computational complexity, so much so that many simplifying assumptions such as wide sense stationarity (WSS) and availability of second order image statistics have

FIGURE 7.1. Digital image restoration system.

been made to obtain computationally feasible algorithms. The inverse filter method works only for extremely high signal-to-noise ratio images. The Wiener filter is usually implemented only after a wide sense stationary assumption (WSS) has been made for images. Furthermore, knowledge of the power spectrum or correlation matrix of the undegraded image is required. Oftentimes, additional assumptions regarding boundary conditions are made so that fast orthogonal transforms can be used. The Kalman filter approach can be applied to the nonstationary image but is computationally very intensive. Similar statements can be made for the SVD pseudoinverse filter method. Approaches based on noncausal models such as the noncausal autoregressive or Gauss Markov random field models [CK82, JC86] also make assumptions such as WSS and periodic boundary conditions. It is desirable to develop a restoration algorithm that does not make WSS assumptions and can be implemented in a reasonable time. An artificial neural network system that can perform extremely rapid computations seems to be very attractive for image restoration in particular and image processing and computer vision in general.

In this chapter, we present a neural network algorithm for restoration of gray-level images degraded by noise and by a known spatially invariant blur function, using a simple sum number representation [TG86]. The gray levels of the image are represented by the simple sum of the neuron state variables, which take binary values of 1 or 0. The restoration procedure consists of two stages: estimation of the parameters of the neural network model and reconstruction of images. During the first stage, the parameters are estimated by comparing the energy function of the neural network with the constrained error function. The nonlinear restoration algorithm is then implemented using a dynamic iterative algorithm to minimize the energy function of the neural network. Owing to the model's fault–tolerant nature and computation capability, a high quality image is obtained using this approach. In order to reduce computational complexity, a practical algorithm, which has results equivalent to those of the original one suggested above, is developed under the assumption that the neurons are sequentially visited. We illustrate the usefulness of this approach by using both synthetic and real images degraded by a known spatially invariant blur function with or without noise. We also discuss the problem of choosing boundary values and introduce two methods to reduce the ringing effect. Comparisons with other restoration methods such as the SVD pseudoinverse filter, the minimum mean square error (MMSE) filter and the modified MMSE filter using Gaussian Markov random field model are given using real images. The advantages of the method presented in this chapter are (1) the WSS assumption is not required for the images, (2) the method can be implemented rapidly and (3) it is fault tolerant.

In the above discussion, it is assumed that the interconnection strengths of the neural network are known from the parameters of the image degradation model and the smoothing constraints. The problem of learning the

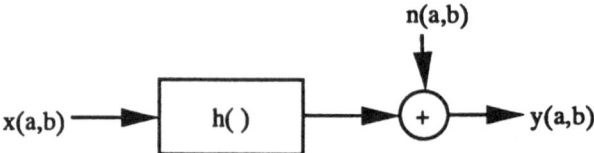

FIGURE 7.2. A typical linear continuous image degradation model.

parameters for the image degradation model from samples of original and degraded images is discussed in [ZCVJ88].

7.2 An Image Degradation Model

The effectiveness of restoration techniques mainly depends on the accuracy of the image modeling. Many image degradation models have been developed based on different assumptions. Figure 7.2 shows a typical linear continuous image degradation model. It is assumed that the image blur can be modeled as a superposition with an impulse response $h(\cdot)$ that may be spatially variant and the output of the system is subject to an additive noise. In this case, the observed image is modeled by

$$y(a,b) = \int_{-\infty}^{\infty} \int_{-\infty}^{\infty} h(a,b;\alpha,\beta)x(\alpha,\beta)d\alpha\,d\beta + n(a,b) \qquad (7.1)$$

where $h(a,b;\alpha,\beta)$ is a blur (or impulse response) function, $n(a,b)$ represents an additive noise, $x(\alpha,\beta)$ and $y(a,b)$ denote the original and degraded (or observed) images, respectively. If the system is spatially invariant, the blur function can be written as

$$h(a,b;\alpha,\beta) = h(a-\alpha,b-\beta).$$

Since many types of degradations can be approximated by a spatially invariant process, our attention will be focussed on the linear spatially invariant model. To measure image quality, the signal to noise ratio (SNR) defined in decibels (dB) is commonly used. The SNR is defined as

$$SNR = 10 \, \log_{10} \frac{\sigma_x^2}{\sigma_n^2}, \qquad (7.2)$$

where σ_x^2 and σ_n^2 are variances of the original image and noise, respectively.

As shown in Figure 7.1, an image digitizer follows the imaging system. The output of the imaging system, the observed image, is first digitized both spatially and in the amplitude and then fed to the restoration system.

For a discrete image restoration system, the object of the restoration is to produce a digital image (a two-dimensional array of samples) that is an estimate of the perfectly digitized original image. Hence, it is necessary to convert the continuous degradation model (7.1) into a discrete one. This can be accomplished by truncating and uniformly sampling the image and blur function to form two–dimensional arrays provided that the sample rate satisfies the Nyquist rate and truncation error is negligibly small. The images are commonly clipped to the intensity range 0 to M, where M is an integer, say $M = 255$.

When a spatially invariant blur function $h(\cdot)$ can be written as a convolution over a small window $K \times K$ (K is an odd integer), the continuous model (7.1) can be written in discrete form as

$$
\begin{aligned}
y(i,j) &= x(i,j) * h(i,j) + n(i,j) \\
&= \sum_{k=-\kappa}^{\kappa} \sum_{l=-\kappa}^{\kappa} x(i-k, j-l)h(k,l) + n(i,j) \qquad (7.3)
\end{aligned}
$$

where indices i and j take the integer number, $*$ denotes the convolution operator and $\kappa = \frac{K-1}{2}$. It is often convenient to express the discrete degradation model (7.3) in terms of vector–matrix form [Win77] as

$$
\underline{Y} = H\underline{X} + \underline{N} \qquad (7.4)
$$

where H is the "blur matrix" corresponding to a blur function, $\underline{N}$ is the signal independent white noise, and $\underline{X}$ and $\underline{Y}$ are the original and degraded images, respectively. This is obtained by row scanning of the arrays. Furthermore, assuming that the dimension of the image array is $L \times L$, H and $\underline{N}$ can be represented as

$$
H = \begin{bmatrix}
h_{1,1} & h_{1,2} & \cdot & \cdot & h_{1,L^2} \\
h_{2,1} & h_{2,2} & \cdot & \cdot & h_{2,L^2} \\
\cdot & \cdot & & & \cdot \\
\cdot & \cdot & & \cdot & \cdot \\
\cdot & \cdot & & & \cdot \\
h_{L^2,1} & h_{L^2,2} & \cdot & \cdot & h_{L^2,L^2}
\end{bmatrix} \qquad (7.5)
$$

and

$$
\underline{N} = \begin{bmatrix} \underline{N}_1 \\ \underline{N}_2 \\ \cdot \\ \cdot \\ \cdot \\ \underline{N}_L \end{bmatrix} = \begin{bmatrix} n_1 \\ n_2 \\ \cdot \\ \cdot \\ \cdot \\ n_{L^2} \end{bmatrix}, \quad \underline{N}_i = \begin{bmatrix} n(i,1) \\ n(i,2) \\ \cdot \\ \cdot \\ \cdot \\ n(i,L) \end{bmatrix} = \begin{bmatrix} n_{(i-1) \times L+1} \\ n_{(i-1) \times L+2} \\ \cdot \\ \cdot \\ \cdot \\ n_{i \times L} \end{bmatrix},
$$

$$(7.6)$$

respectively. Vectors $\underline{X}$ and $\underline{Y}$ have representations similar to that of $\underline{N}$. Note that this expression is derived under the assumption of a linear, spatially invariant degradation model. Equation (7.4) is similar to the simultaneous equations solution of [TG86], but differs in that it includes a noise term.

For instance, if the blur function takes the form

$$h(k, l) = \begin{cases} 0.5 & if \ k = 0, \ l = 0 \\ 0.0625 & if \ |k|, \ |l| \leq 1, \ (k, l) \neq (0, 0) \end{cases} \tag{7.7}$$

then the "blur matrix" H will be a block circulant or block Toeplitz matrix according to the boundary conditions. The general form of H corresponding to (7.7) is as follows:

$$H = \begin{bmatrix} H_0 & H_1 & \underline{0} & \cdot & \cdot & \cdot & \underline{0} & H_c \\ H_1 & H_0 & H_1 & \cdot & \cdot & \cdot & \underline{0} & \underline{0} \\ \cdot & \cdot & & & & & \cdot & \cdot \\ \cdot & \cdot & & & & & & \\ \cdot & \cdot & \cdot & & & & \cdot & \cdot \\ H_c & \underline{0} & \underline{0} & \cdot & \cdot & \cdot & H_1 & H_0 \end{bmatrix} \tag{7.8}$$

where

$$H_0 = \begin{bmatrix} \frac{1}{2} & \frac{1}{16} & 0 & \cdot & \cdot & \cdot & 0 & h_c \\ \frac{1}{16} & \frac{1}{2} & \frac{1}{16} & \cdot & \cdot & \cdot & 0 & 0 \\ \cdot & \cdot & \cdot & & & & \cdot & \cdot \\ \cdot & \cdot & \cdot & & & & & \\ \cdot & \cdot & \cdot & & & & \cdot & \cdot \\ h_c & 0 & 0 & \cdot & \cdot & \cdot & \frac{1}{16} & \frac{1}{2} \end{bmatrix} \tag{7.9}$$

$$H_1 = \begin{bmatrix} \frac{1}{16} & \frac{1}{16} & 0 & \cdot & \cdot & \cdot & 0 & h_c \\ \frac{1}{16} & \frac{1}{16} & \frac{1}{16} & \cdot & \cdot & \cdot & 0 & 0 \\ \cdot & \cdot & \cdot & & & & \cdot & \cdot \\ \cdot & \cdot & \cdot & & & & & \\ \cdot & \cdot & \cdot & & & & \cdot & \cdot \\ h_c & 0 & 0 & \cdot & \cdot & \cdot & \frac{1}{16} & \frac{1}{16} \end{bmatrix} \tag{7.10}$$

and $\underline{0}$ is the null matrix whose elements are all zeros. If the image has periodic boundaries, H becomes a block circulant matrix, thus $h_c = 0.0625$ and $H_c = H_1$. If the image boundaries are padded by zeros, H is then a block Toeplitz matrix with $h_c = 0$ and $H_c = \underline{0}$.

7.3   Image Representation

We use a neural network containing redundant neurons for representing the image gray levels. The model consists of $L^2 \times M$ mutually interconnected neurons, where L is the size of image and M is the maximum value of the gray level function. The image is described by a finite set of gray level functions $\{x(i,j), \ where \ 1 \le i,j \le L\}$ with $x(i,j)$ (positive integer number) denoting the gray level of the pixel (i,j). Let $V = \{v_{i,k}, \ where \ 1 \le i \le L^2, 1 \le k \le M\}$ be a binary state set of the neural network with $v_{i,k}$ (1 for firing and 0 for resting) denoting the state of the (i,k)th neuron. The image gray level function is represented by a simple sum of the neuron state variables as

$$x(i,j) = \sum_{k=1}^{M} v_{m,k} \qquad (7.11)$$

where $m = (i-1) \times L + j$. Here the gray level functions have degenerate representations. For instance, if a gray level function is represented by M neurons and it takes a value of 10, then any 10 out of M neurons will fire and there are $\frac{M!}{10!(M-10)!}$ representations. This is a one-to-many mapping from an integer number space to a neuron state space. Note that any single misfiring neuron does not cause a large error in the number representation. It is interesting to see that the network using such a representation scheme has $\prod_{i=1}^{L} \prod_{i=1}^{L} \frac{M!}{x(i,j)!(M-x(i,j))!}$ stable states for an $L \times L$ image! So many stable states offer the network more chances to reach a correct solution. Hence, use of this redundant number representation scheme yields advantages such as fault–tolerance and faster convergence to a solution [TG86].

In this model, each neuron (i,k) randomly and asynchronously receives inputs from all neurons and a bias input

$$u_{i,k} = \sum_{j}^{L^2} \sum_{l}^{M} T_{i,k;j,l} v_{j,l} + I_{i,k} \qquad (7.12)$$

where $T_{i,k;j,l}$ denotes the strength (possibly negative) of the interconnection between neuron (i,k) and neuron (j,l) and $I_{i,k}$ is a bias input. We assume that the interconnection strength has the following properties:

$$T_{i,k;j,l} = T_{j,l;i,k}$$

and

$$T_{i,k;i,k} \ne 0,$$

which means that the strengths are symmetric and neurons have self–feedback. Each $u_{i,k}$ is fed back to corresponding neurons after thresholding

$$v_{i,k} = g(u_{i,k}) \qquad (7.13)$$

where $g(x)$ is a nonlinear function whose form can be taken as

$$g(x) = \begin{cases} 1 & if \ x \ge 0 \\ 0 & if \ x < 0. \end{cases} \qquad (7.14)$$

In this model, the state of each neuron is updated by using the latest information about other neurons.

7.4 Estimation of Model Parameters

The neural model parameters, the interconnection strengths and bias inputs, can be determined in terms of the energy function of the neural network. As defined in [HT85], the energy function of the neural network can be written as

$$E = -\frac{1}{2} \sum_{i=1}^{L^2} \sum_{j=1}^{L^2} \sum_{k=1}^{M} \sum_{l=1}^{M} T_{i,k;j,l}\, v_{i,k}\, v_{j,l} - \sum_{i=1}^{L^2} \sum_{k=1}^{M} I_{i,k}\, v_{i,k} \qquad (7.15)$$

In order to use the spontaneous energy–minimization process of the neural network, we reformulate the restoration problem as one of minimizing an error function with constraints defined as

$$E = \frac{1}{2}\|\underline{Y} - H\underline{\hat{X}}\|^2 + \frac{1}{2}\lambda\|D\underline{\hat{X}}\|^2 \qquad (7.16)$$

where $\|\underline{Z}\|$ is the L_2 norm of $\underline{Z}$ and λ is a constant. Such a constrained error function is widely used in image restoration problems [AH77] and is also similar to the regularization techniques used in early vision problems [PTK85]. The first term in (7.16) is to seek an $\underline{\hat{X}}$ such that $H\underline{\hat{X}}$ approximates $\underline{Y}$ in a least squares sense. Meanwhile, the second term is a smoothness constraint on the solution $\underline{\hat{X}}$. The constant λ determines their relative importance to achieve suppression of noise and ringing. .

In general, if H is a low pass distortion, then D is a high pass filter. A common choice of D is a second order differential operator which can be approximated as a local window operator in the two-dimensional discrete case. For instance, if D is a Laplacian operator

$$\nabla = \frac{\partial^2}{\partial i^2} + \frac{\partial^2}{\partial j^2} \qquad (7.17)$$

it can be approximated as a window operator

$$\frac{1}{6}\begin{bmatrix} 1 & 4 & 1 \\ 4 & -20 & 4 \\ 1 & 4 & 1 \end{bmatrix}. \qquad (7.18)$$

Then D will be a block Toeplitz matrix similar to (7.8).

Expanding (7.16) and then replacing x_i by (7.11), we have

$$\begin{aligned} E &= \frac{1}{2}\sum_{p=1}^{L^2}\left(y_p - \sum_{i=1}^{L^2} h_{p,i} x_i\right)^2 + \frac{1}{2}\lambda\sum_{p=1}^{L^2}\left(\sum_{i=1}^{L^2} d_{p,i} x_i\right)^2 \\ &= \frac{1}{2}\sum_{i=1}^{L^2}\sum_{j=1}^{L^2}\sum_{k=1}^{M}\sum_{l=1}^{M}\sum_{p=1}^{L^2} h_{p,i}\, h_{p,j}\, v_{i,k}\, v_{j,l} + \frac{1}{2}\sum_{p=1}^{L^2} y_p^2 \end{aligned}$$

$$-\sum_{i=1}^{L^2} \sum_{k=1}^{M} \sum_{p=1}^{L^2} y_p\, h_{p,i}\, v_{i,k}$$

$$+\frac{1}{2}\lambda \sum_{i=1}^{L^2} \sum_{j=1}^{L^2} \sum_{k=1}^{M} \sum_{l=1}^{M} \sum_{p=1}^{L^2} d_{p,i}\, d_{p,j}\, v_{i,k}\, v_{j,l}. \tag{7.19}$$

By comparing the terms in (7.19) with the corresponding terms in (7.15) and ignoring the constant term $\frac{1}{2}\sum_{p=1}^{L^2} y_p^2$, we can determine the interconnection strengths and bias inputs as

$$T_{i,k;j,l} = -\sum_{p=1}^{L^2} h_{p,i}\, h_{p,j} - \lambda \sum_{p=1}^{L^2} d_{p,i}\, d_{p,j} \tag{7.20}$$

and

$$I_{i,k} = \sum_{p=1}^{L^2} y_p\, h_{p,i} \tag{7.21}$$

where $h_{i,j}$ and $d_{i,j}$ are the elements of the matrices H and D, respectively. Two interesting aspects of (7.20) and (7.21) should be pointed out: (1) the interconnection strengths are independent of subscripts k and l and the bias inputs are independent of subscript k, and (2) the self-connection $T_{i,k;i,k}$ is not equal to zero which requires self-feedback for neurons.

From (7.20), one can see that the interconnection strengths are determined by the shift–invariant blur function, differential operator and constant λ. Hence, $T_{i,k;j,l}$ can be computed without error provided the blur function is known. However, the bias inputs are functions of the observed degraded image. If the image is degraded by a shift–invariant blur function only, then $I_{i,k}$ can be estimated perfectly. Otherwise, $I_{i,k}$ is affected by noise. The reasoning behind this statement is as follows. By replacing y_p by $\sum_{i=1}^{L^2} h_{p,i}\, x_i + n_p$, we have

$$\begin{aligned} I_{i,k} &= \sum_{p=1}^{L^2} (\sum_{i=1}^{L^2} h_{p,i}\, x_i + n_p)\, h_{p,i} \\ &= \sum_{p=1}^{L^2} \sum_{i=1}^{L^2} h_{p,i}\, x_i\, h_p + \sum_{p=1}^{L^2} n_p\, h_{p,i}. \end{aligned} \tag{7.22}$$

The second term in (7.22) represents the effects of noise. If the signal to noise ratio (SNR) is low, then we have to choose a large λ to suppress effects due to noise. It seems that in the absence of noise, the parameters can be estimated perfectly, ensuring exact recovery of the image as the error function E tends to zero. However, the problem is not so simple, since the restoration performance depends on both the parameters and the blur function when a mean square error or least square error such as (7.16) is used. A discussion about the effects of blur function is given in Section 7.11.

7.5   Restoration

Restoration is carried out by neuron evaluation and an image construction procedure. Once the parameters $T_{i,k;j,l}$ and $I_{i,k}$ are obtained using (7.20) and (7.21), each neuron can randomly and asynchronously evaluate its state and readjust accordingly using (7.12) and (7.13). When one quasi–minimum energy point is reached, the image can be constructed using (7.11).

Although a step function is used as an activation function, the self-feedback may still cause the energy function E to increase with a transition. This is explained below. Define the state change $\Delta v_{i,k}$ of neuron (i,k) and energy change ΔE as

$$\Delta v_{i,k} = v_{i,k}^{new} - v_{i,k}^{old} \quad and \quad \Delta E = E^{new} - E^{old}.$$

Consider the energy function

$$E = -\frac{1}{2} \sum_{i=1}^{L^2} \sum_{j=1}^{L^2} \sum_{k=1}^{M} \sum_{l=1}^{M} T_{i,k;j,l}\, v_{i,k}\, v_{j,l} - \sum_{i=1}^{L^2} \sum_{k=1}^{M} I_{i,k}\, v_{i,k}. \qquad (7.23)$$

Since only one neuron is updated at each step, the energy change ΔE due to a change $\Delta v_{i,k}$ is given by

$$\Delta E = -\left(\sum_{j=1}^{L^2} \sum_{l=1}^{M} T_{i,k;j,l}\, v_{j,l} + I_{i,k}\right)\Delta v_{i,k} - \frac{1}{2} T_{i,k;i,k}\, (\Delta v_{i,k})^2, \qquad (7.24)$$

which is not always negative. For instance, if

$$v_{i,k}^{old} = 0, \quad u_{i,k} = \sum_{j=1}^{L^2} \sum_{l=1}^{M} T_{i,k;j,l}\, v_{j,l} + I_{i,k} > 0,$$

and the threshold function is as in (7.14), then $v_{i,k}^{new} = 1$ and $\Delta v_{i,k} > 0$. Thus, the first term in (7.24) is negative. But

$$T_{i,k;i,k} = -\sum_{p=1}^{L^2} h_{p,i}^2 - \lambda \sum_{p=1}^{L^2} d_{p,i}^2 < 0$$

with $\lambda > 0$ leading to

$$-\frac{1}{2} T_{i,k;i,k}\, (\Delta v_{i,k})^2 > 0.$$

When the first term is less than the second term in (7.24), then $\Delta E > 0$.

Thus, depending on whether convergence to a local minimum or a global minimum is desired, we can use a deterministic or stochastic decision rule. The restoration algorithm is summarized below.

Algorithm 1:

1. Set the initial state of the neurons.

2. Update the state of all neurons randomly and asynchronously according to the decision rule.

3. Check the energy function; if the energy does not change, go to Step 4; otherwise, go back to Step 2.

4. Construct an image using (7.11).

7.6   A Practical Algorithm

The algorithm described above is difficult to simulate on a conventional computer owing to high computational complexity even for images of reasonable size. For instance, if we have an $L \times L$ image with M gray levels, then $L^2 M$ neurons and $\frac{1}{2} L^4 M^2$ interconnections are required and $L^4 M^2$ additions and multiplications are needed at each iteration. Therefore, the space and time complexities are $O(L^4 M^2)$ and $O(L^4 M^2 K)$, respectively, where K, typically $10 - 100$, is the number of iterations. Usually, L and M are $256 - 1024$ and 256, respectively. However, simplification is possible if the neurons are sequentially updated.

In order to simplify the algorithm, we begin by reconsidering equations (7.12) and (7.13). As noted earlier, the interconnection strengths given in (7.20) are independent of subscripts k and l and the bias inputs given in (7.21) are independent of subscript k, so the M neurons used to represent the same image gray level function have the same interconnection strengths and bias inputs. Hence, one set of interconnection strengths and one bias input are sufficient for every gray level function; thus, the dimensions of the interconnection strength matrix T and bias input matrix I can be reduced by a factor of M^2. From (7.12) all inputs received by a neuron, for example, the (i, k)th neuron can be written as

$$
\begin{aligned}
u_{i,k} &= \sum_{j}^{L^2} T_{i,\cdot;j,\cdot} \left(\sum_{l}^{M} v_{j,l} \right) + I_{i,\cdot} \\
&= \sum_{j}^{L^2} T_{i,\cdot;j,\cdot}\, x_j + I_{i,\cdot}
\end{aligned}
\qquad (7.25)
$$

where we have used (7.11), and x_j is the gray level function of the jth image pixel. The symbol "$\cdot$" in the subscripts means that the $T_{i,\cdot;j,\cdot}$ and $I_{i,\cdot}$ are independent of k. Equation (7.25) suggests that we can use a multi-value number to replace the simple sum number. Since the interconnection strengths are determined by the blur function, the differential operator and the constant λ as shown in (7.20), it is easy to see that if the blur function is local, then most interconnection strengths are zeros and the neurons are locally connected. Therefore, most elements of the interconnection matrix T are zeros. If the blur function is shift invariant taking the form in (7.7), then the interconnection matrix is block Toeplitz so that only a few elements need to be stored. Based on the value of inputs $u_{i,k}$, the state of the (i,k)th neuron is updated by applying a decision rule. The state change of the (i,k)th neuron in turn causes the gray level function x_i to change as

$$x_i^{new} = \begin{cases} x_i^{old} & if \ \Delta v_{i,k} = 0 \\ x_i^{old} + 1 & if \ \Delta v_{i,k} = 1 \\ x_i^{old} - 1 & if \ \Delta v_{i,k} = -1 \end{cases} \tag{7.26}$$

where $\Delta v_{i,k} = v_{i,k}^{new} - v_{i,k}^{old}$ is the state change of the (i,k)th neuron. The superscripts "new" and "old" are for after and before updating, respectively. We use x_i to represent the gray level value as well as the output of M neurons representing x_i. Assuming that the neurons of the network are sequentially visited, it is straightforward to show that the updating procedure can be reformulated as

$$u_{i,k} = \sum_j^{L^2} T_{i,\cdot;j,\cdot} \, x_j + I_{i,\cdot}. \tag{7.27}$$

$$\Delta v_{i,k} = g(u_{i,k}) = \begin{cases} \Delta v_{i,k} = 0 & if \ u_{i,k} = 0 \\ \Delta v_{i,k} = 1 & if \ u_{i,k} > 0 \\ \Delta v_{i,k} = -1 & if \ u_{i,k} < 0 \end{cases} \tag{7.28}$$

$$x_i^{new} = \begin{cases} x_i^{old} + \Delta v_{i,k} & if \ \Delta E < 0 \\ x_i^{old} & if \ \Delta E \geq 0. \end{cases} \tag{7.29}$$

Note that the stochastic decision rule can also be used in (7.29). In order to limit the gray level function to the range 0–255 after each updating step, we have to check the value of the gray level function x_i^{new}. Equations (7.27), (7.28), and (7.29) give a much simpler algorithm. This algorithm is summarized below.

Algorithm 2:

1. Take the degraded image as the initial value.

2. Sequentially visit all numbers (image pixels). For each number, use (7.27), (7.28), and (7.29) to update it repeatedly until no further

change; thus, if $\Delta v_{i,k} = 0$ or energy change $\Delta E \geq 0$, then move to the next one.

3. Check the energy function; if the energy does not change anymore, a restored image is obtained; otherwise, go back to Step 2 for another iteration.

The calculations of the inputs $u_{i,k}$ of the (i,k)th neuron and the energy change ΔE can be simplified further. When we update the same image gray level function repeatedly, the input received by the current neuron (i,k) can be computed by making use of the previous result

$$u_{i,k} = u_{i,k-1} + \Delta v_{i,k}\, T_{i,.;i,.} \tag{7.30}$$

where $u_{i,k-1}$ is the input received by the $(i,k-1)$th neuron. The energy change ΔE due to the state change of the (i,k)th neuron can be calculated as

$$\Delta E = -u_{i,k}\,\Delta v_{i,k} - \frac{1}{2}\, T_{i,.;i,.}\, (\Delta v_{i,k})^2. \tag{7.31}$$

If the blur function is shift invariant, all these simplifications reduce the space and time complexities significantly from $O(L^4 M^2)$ and $O(L^4 M^2 K)$ to $O(L^2)$ and $O(ML^2 K)$, respectively. Since every gray level function needs only a few updating steps after the first iteration, the computation at each iteration is $O(L^2)$. The resulting algorithm can be easily simulated on mini-computers for images as large as 512×512.

7.7 Computer Simulations

The practical algorithm described in the previous section was applied to synthetic and real images on a Sun-3/160 Workstation. In all cases only the deterministic decision rule was used. The results are summarized in Figures 7.3 and 7.4.

Figure 7.3 shows the results for a synthetic image. The original image shown in Figure 7.3(a) is of size 32×32 with three gray levels. The image was degraded by convolving with a 3×3 blur function as in (7.7) using circulant boundary conditions; 22 dB white Gaussian noise was added after convolution. A perfect image was obtained after 6 iterations without preprocessing. We set the initial state of all neurons to equal 1, meaning they were firing and chose $\lambda = 0$ due to the good conditioning of the blur function.

Figure 7.4(a) shows the original girl image and Figure 7.4(b) shows the noisy blurred image. The original image is of size 256×256 with 256 gray levels. The variance of the original image is 2797.141. It was degraded by a 5×5 uniform blur function. A small amount of quantization noise was introduced by quantizing the convolution results to eight bits. For

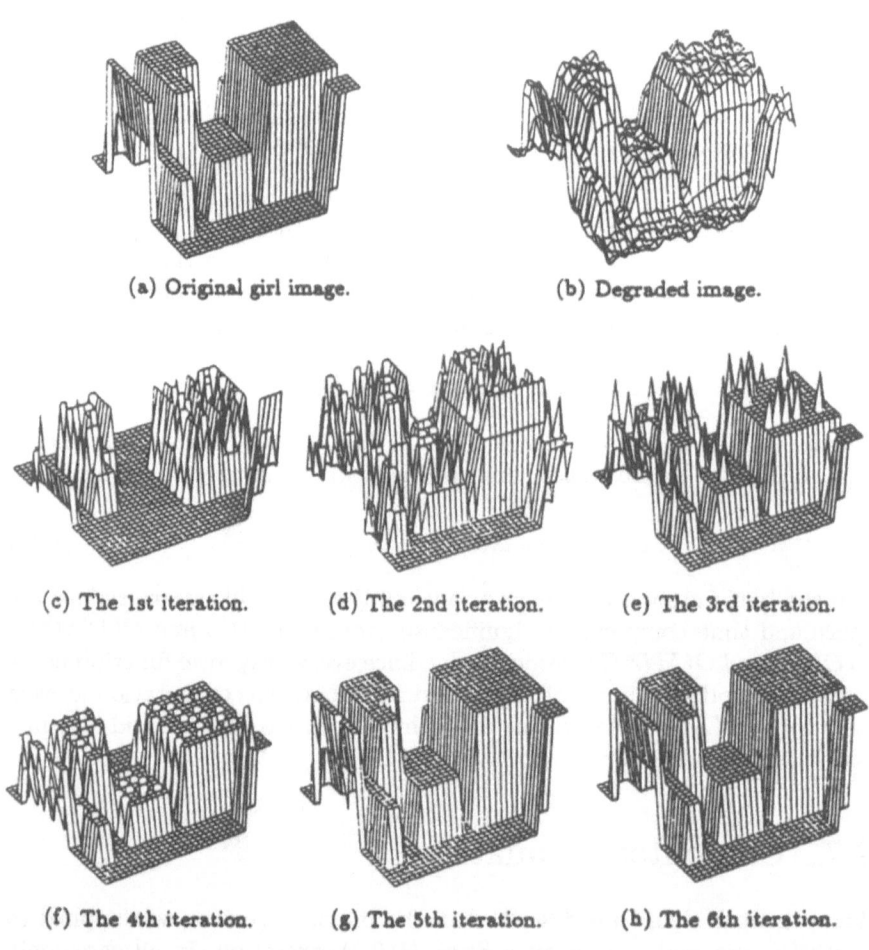

(a) Original girl image.   (b) Degraded image.

(c) The 1st iteration.   (d) The 2nd iteration.   (e) The 3rd iteration.

(f) The 4th iteration.   (g) The 5th iteration.   (h) The 6th iteration.

FIGURE 7.3. Restoration of noisy blurred synthetic image.

comparison purposes, Figure 7.4(c) shows the outputs of an inverse filter [PFG78] and Figure 7.4(d) shows the outputs of our approach. The restored image given by the inverse filter is completely overridden by the amplified noise and the ringing effects due to the ill conditioned blur matrix H. Since the blur matrix H corresponding to the 5×5 uniform blur function is not singular, the pseudoinverse filter [PFG78] and the inverse filter have the same output. However, the problem of the restored image overridden by noise does not occur with our approach. In order to avoid the ringing effects due to the boundary conditions, we took four pixel wide boundaries, the first and last four rows and columns, from the original image, and updated the interior region (248×248) of the image only. The noisy blurred image was used as an initial condition for accelerating the convergence. The constant λ was set at 0 because of small noise and good boundary values. The restored image was obtained after 213 iterations. The square error (the energy function) defined in (7.16) is 0.02543 and the square error between the original and the restored image is 66.5027.

7.8   Choosing Boundary Values

As mentioned in [WBT85], choosing boundary values is a common problem for techniques ranging from deterministic inverse filter algorithms to stochastic Kalman filters. In these algorithms boundary values determine the entire solution when the blur is uniform [Son72]. The same problem occurs in the neural network approach. Since the 5×5 uniform blur function is ill conditioned, improper boundary values may cause ringing which may affect the restored image completely. For example, appending zeros to the image as boundary values introduces a sharp edge at the image border and triggers ringing in the restored image even if the image has zero mean. Another procedure is to assume a periodic boundary. When the left (top) and right (bottom) borders of the image are different, a sharp edge is formed and ringing results even though the degraded image has been formed by blurring with periodic boundary conditions. The drawbacks of these two assumptions for boundary values were reported in [WBT85, WI81, BRG83] for the two-dimensional Kalman filtering technique. We also tested our algorithm using these two assumptions for boundary values; the results indicate the restored images were seriously affected by ringing.

In the last section, to avoid the ringing effect we took 4 pixels wide borders from the original image as boundary values for restoration. Since the original image is not available in practice, an alternative to eliminating the ringing effect caused by sharp false edges is to use the blurred noisy boundaries from the degraded image. Figure 7.5(a) shows the restored image using the first and last four rows and columns of the blurred noisy image in Figure 7.4 as boundary values. In the restored image there still exists some ringing due to the naturally occurring sharp edges in the region

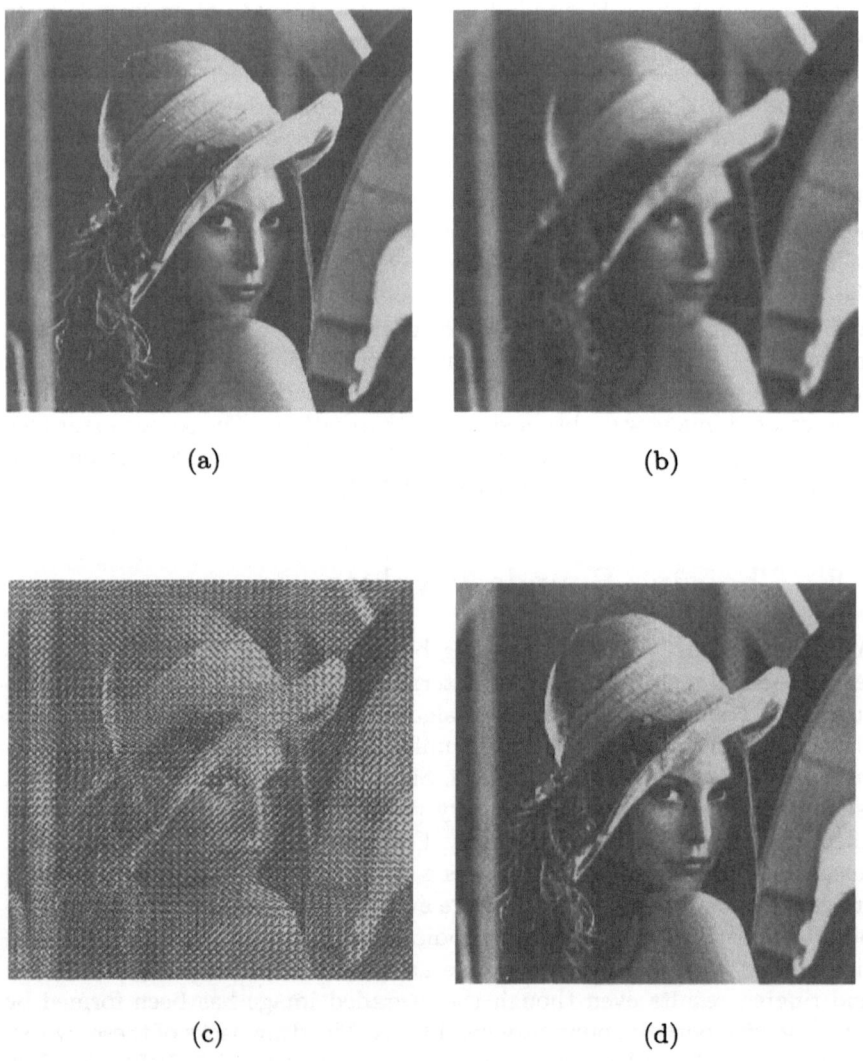

(a)   (b)

(c)   (d)

FIGURE 7.4. Restoration of noisy blurred real image. (a) Original girl image. (b) Image degraded by 5×5 uniform blur and quantization noise. (c) Restored image using inverse filter. (d) Restored image using our approach. (Reprinted with permission from [ZCVJ88] Figure 2, ©1988 IEEE.)

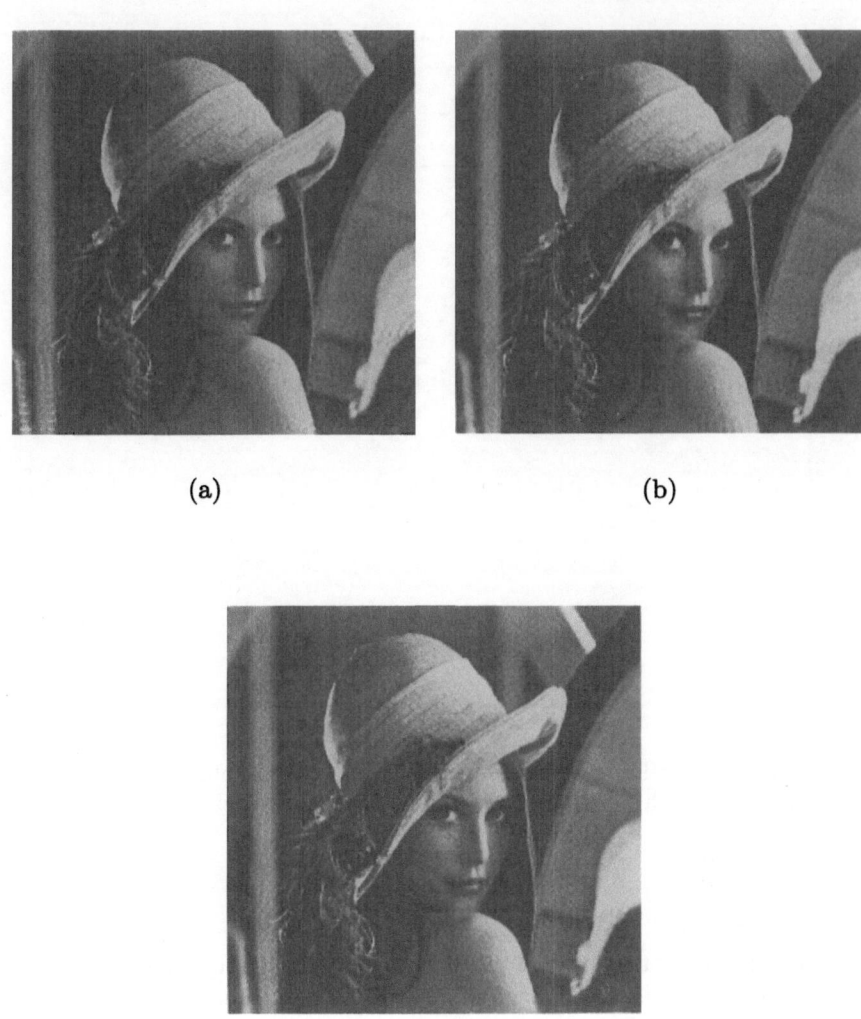

(a)

(b)

(c)

FIGURE 7.5. Restored images using different boundary conditions. (a) Blurred noisy boundaries. (b) Method 1. (c) Method 2. (Reprinted with permission from [ZCVJ88] Figure 3, ©1988 IEEE.)

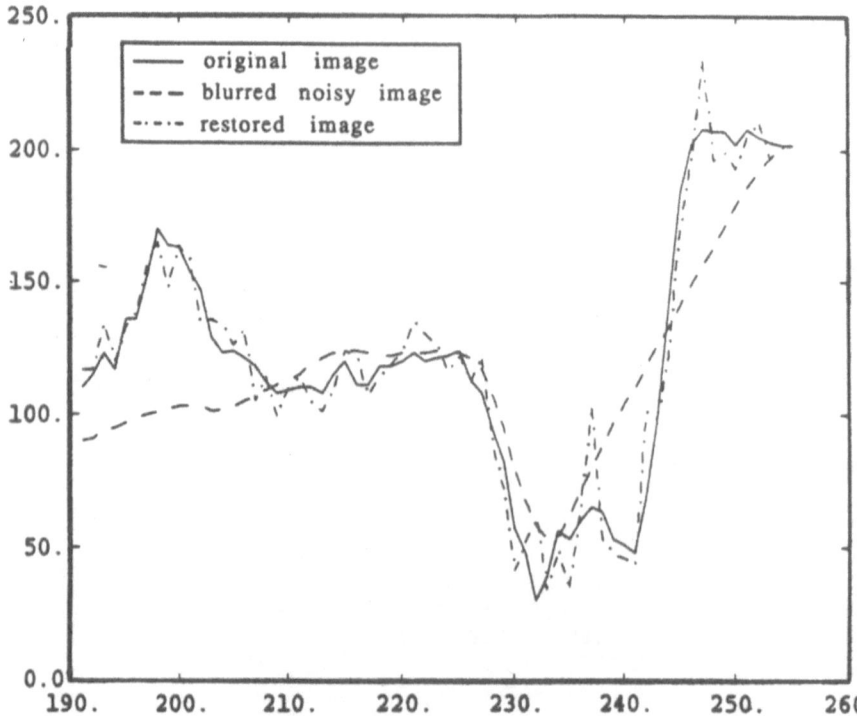

FIGURE 7.6. One typical cut of the restored image using the blurred noisy boundaries. (Reprinted with permission from [ZCVJ88] Figure 4, ©1988 IEEE.)

near the borders in the original image, but not due to boundary values. A typical cut of the restored image to illustrate ringing near the borders is shown in Figure 7.6. To remove the ringing near the borders caused by naturally occurring sharp edges in the original image, we suggest the following techniques.

First, divide the image into three regions: border, subborder and interior region as shown in Figure 7.7. For the 5×5 uniform blur case, the border region will be four pixels wide due to the boundary effect of the bias input $I_{i,k}$ in (7.21), and the subborder region will be four or eight pixels wide. In fact, the width of the subborder region will be image-dependent. If the regions near the border are smooth, then the width of the subborder region will be small or even zero. If the border contains many sharp edges the width will be large. For the real girl image, we chose the width of the subborder region to be eight pixels. We suggest using one of the following two methods.

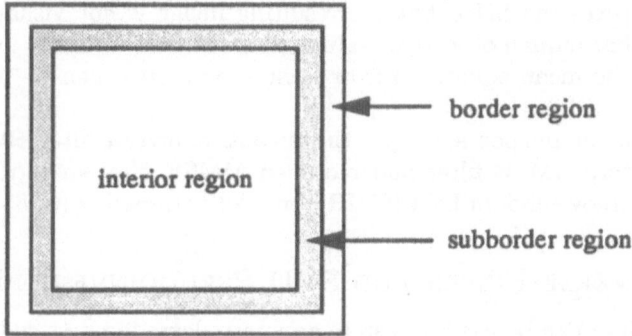

FIGURE 7.7. Border, subborder, and interior regions of the image.

Method 1: In the case of small noise, such as quantization error noise, the blurred image is usually smooth. Therefore, we restricted the difference between the restored and blurred image in the subborder region to a certain range to reduce the ringing effect. Mathematically, this constraint can be written as

$$\|\hat{x}_i - y_i\| \leq T \quad \text{for } i \in \text{ subborder region,} \qquad (7.32)$$

where T is a threshold and $\hat{x}_i$ is the restored image gray value.

Method 2: This method simply sets λ in (7.16) to zero in the interior region and nonzero in the subborder region, respectively.

Figure 7.5(b) shows the results of using Method 1 and Figure 7.5(c) shows the results of using Method 2. For Method 1, T is set at 10. For Method 2, λ is set at 0.09 and D is a Laplacian operator. Since it checks all restored image gray values in the subborder region, Method 1 needs more computation than Method 2. However, Method 2 is very sensitive to the parameter λ while Method 1 is not so sensitive to the parameter λ. Experimental results show that both Methods 1 and 2 reduce the ringing effect significantly by using the suboptimal blurred boundary values.

7.9   Comparisons to Other Restoration Methods

Comparison of the performance of different restoration methods requires some quality measures which are difficult to define due to the lack of knowledge about the human visual system. The word "optimal" used in the restoration techniques usually refers only to a mathematical concept, and is not related to the response of the human visual system. For instance, when the blur function is ill conditioned and the SNR is low, the MMSE method improves the SNR, but the resulting image is not visually good. We believe that human objective evaluation is the best ultimate judgment. Meanwhile, the mean square error or least square error can be used as a reference.

For comparison purposes, we give the outputs of inverse filter, SVD pseudoinverse filter, MMSE filter and modified MMSE filter in terms of the Gaussian Markov random field (GMRF) model parameters [CJ85, JC86].

7.9.1   INVERSE FILTER AND SVD PSEUDOINVERSE FILTER

An inverse filter can be used to restore an image degraded by a spatially invariant blur function with high signal to noise ratio. When the blur function has some singular points, a SVD pseudoinverse filter is needed; however, both filters are very sensitive to noise. This is because for restoration purposes the noise is amplified in the same way as are the signal components. The inverse filter and SVD pseudoinverse filter were applied to an image degraded by the 5×5 uniform blur function and quantization noise (approximately 40 dB SNR). The blurred and restored images are shown in Figure 7.4. As we mentioned before the outputs of these filters are completely overridden by the amplified noise and ringing effects.

7.9.2   MMSE AND MODIFIED MMSE FILTERS

The MMSE filter is also known as the Wiener filter (in frequency domain). Under the assumption that the original image obeys a GMRF model, the MMSE filter (or Wiener filter) can be represented in terms of the GMRF model parameters and the blur function. In our implementation of the MMSE filter, we used a known blur function, unknown noise variance and the GMRF model parameters estimated from the blurred noisy image by a maximum likelihood (ML) method [CJ85]. The image shown in Figure 7.8(a) was degraded by 5×5 uniform blur function and 20 dB SNR additive white Gaussian noise. The restored image using the MMSE filter is shown in Figure 7.8(b).

The modified MMSE filter in terms of the GMRF model parameters is a linear weighted combination of a Wiener filter with a smoothing operator (such as median filter) and a pseudoinverse filter to smooth the noise and preserve the edge of the restored image simultaneously. Details of this filter

TABLE 7.1. Mean square error (MSE) improvement.

MMSE	MMSE (o)	Modified MMSE	Neural network
1.384 dB	2.139 dB	1.893 dB	1.682 dB

can be found in [JC86]. We applied the modified MMSE filter to the same image used in the MMSE filter above with the same model parameters. The smoothing operator is a 9×9 cross shape median filter. The resulting image using the modified MMSE filter is also shown in Figure 7.8(c). The result of our method is shown in Figure 7.8(d). The D we used in (7.16) is a Laplacian operator as in (7.17). We chose $\lambda = 0.0625$ and used blurred noisy boundaries, that are four pixels wide, for restoration. The total number of iterations was 20. The improvement of mean square error between the restored image and the original image for each method is shown in Table 7.1. In the table the "MMSE (o)" denotes that the parameters were estimated from the original image. The restored image using the "MMSE (o)" is very similar to Figure 7.8(a). As we mentioned before, the comparison of the outputs of the different restoration methods is a difficult problem. The MMSE filter visually gives the worst output which has the smallest mean square error for the MMSE(o) case. The result of our method is smoother than that of the MMSE filter. Although the output of the modified MMSE filter is smooth in flat regions, it contains some artifacts and snake effects at the edges, a consequence of using a large sized median filter.

7.10   Optical Implementation

In order to take advantage of the parallelism of the optics, we propose to use a semisynchronous neural network instead of an asynchronous one. The difference between the semisynchronous and asynchronous neural networks is that at each clock cycle the former updates L^2 neurons selected from L^2 different gray level functions, whereas the latter updates one neuron only. Without loss of generality, we assume that the neurons are semisequentially updated which is not similar to natural neuron transition rules. The semisequential updating means that if an $M \times L^2$ matrix is used to represent neurons (each element for one neuron and each column for one gray level function), then each row will be sequentially updated.

A system capable of performing the semisynchronous neural network is shown in Figure 7.9. The network consists of the equations (7.12) and (7.13) with a deterministic decision rule. The system is based on the idea described in [FPPP85] and [AMP87] using an optical matrix-matrix product. The basic idea is as follows.

As noted earlier, the matrix-matrix product (7.12) can be written as

$$u_{i,k} = \sum_{j}^{L^2} T_{i,\cdot;j,\cdot} \left(\sum_{l}^{M} v_{j,l} \right) + I_{i,}, \qquad (7.33)$$

which is a vector-matrix product. We use an $M^2 \times (L^2 + 1)$ laser diode array shown in Figure 7.10 to represent the current state of neurons stored

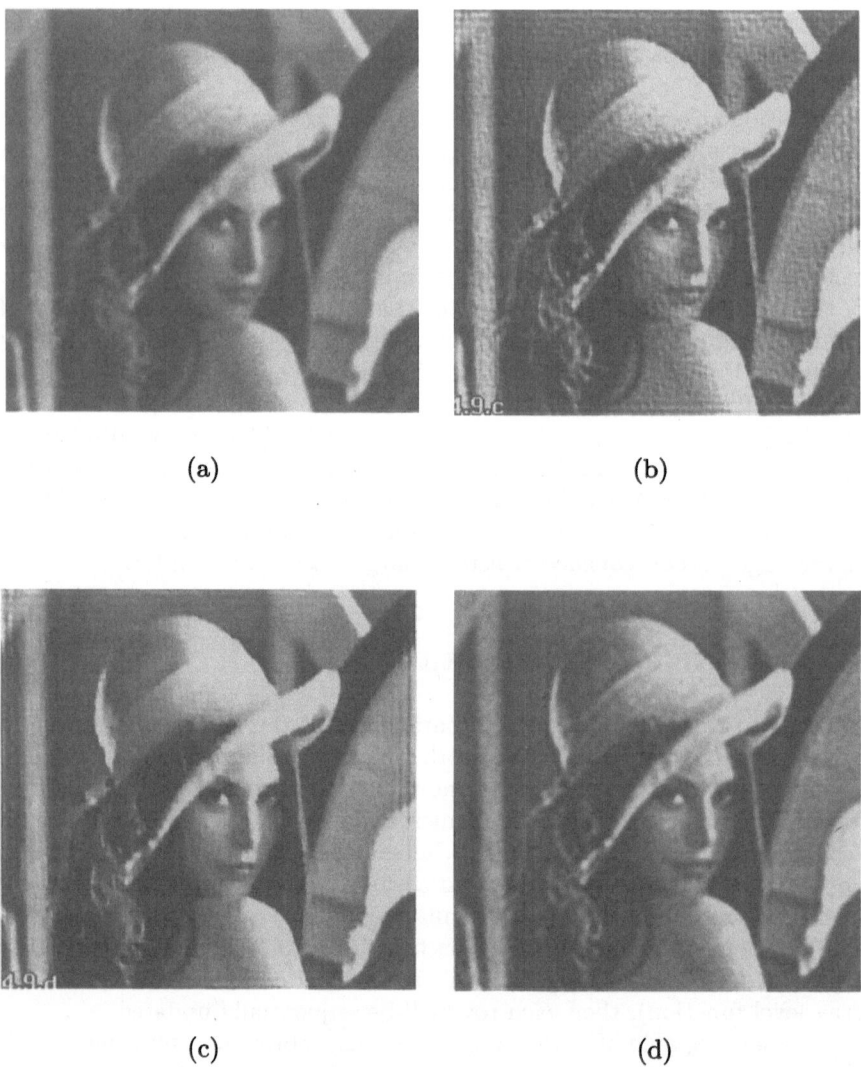

(a)   (b)

(c)   (d)

FIGURE 7.8. Comparision to other restoration methods. (a) Image degraded by 5 × 5 uniform blur and 20 dB SNR additive white Gaussian noise. (b) Restored image using the MMSE filter. (c) Restored image using the modified MMSE filter. (d) Restored image using our approach. (Reprinted with permission from [ZCVJ88] Figure 6, ©1988 IEEE.)

LED ◄─ OPTICS ─► SLM ◄─ OPTICS ─► PD

V   T, I   U

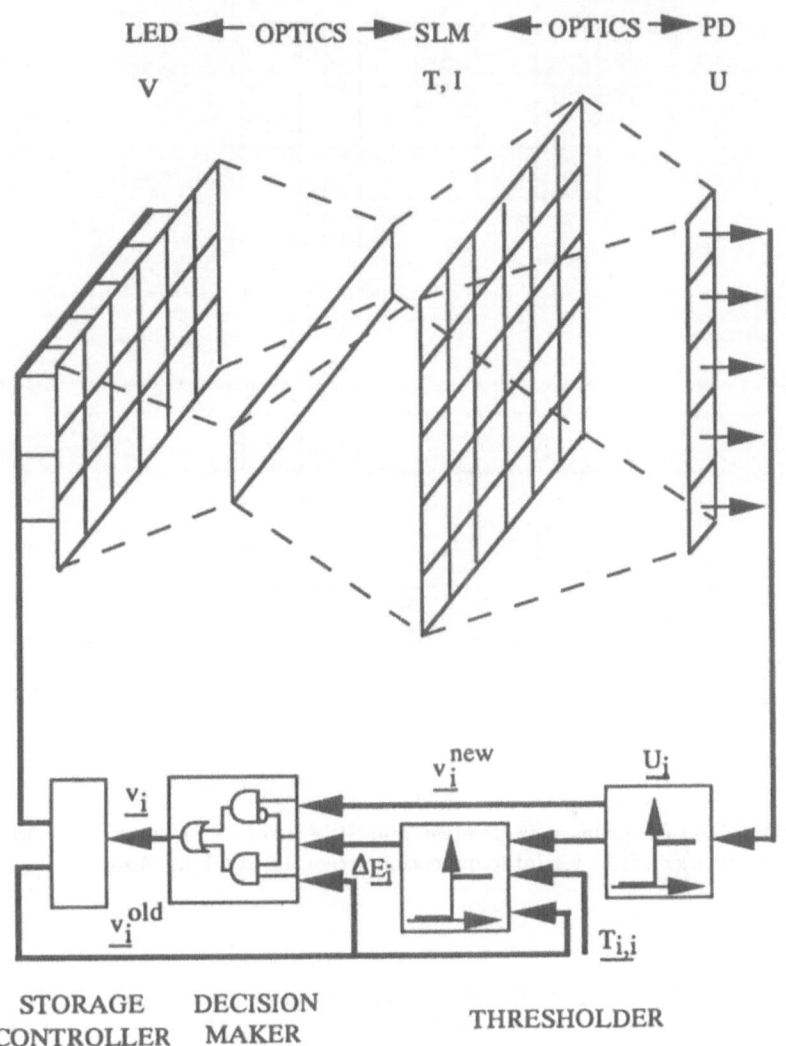

STORAGE   DECISION
CONTROLLER MAKER   THRESHOLDER

FIGURE 7.9. A schematic diagram for implementing the semisynchronous neural network.

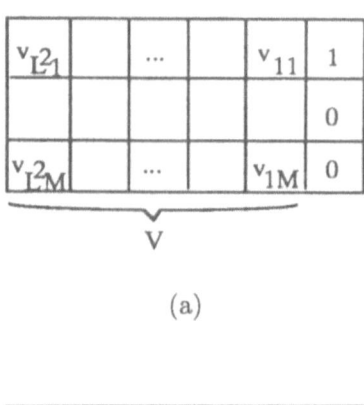

(a)

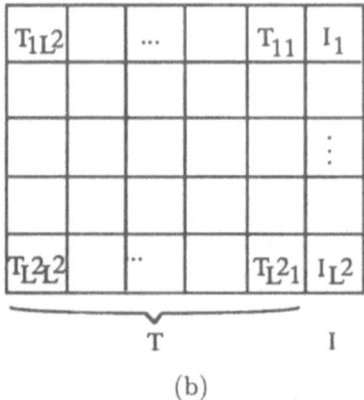

(b)

FIGURE 7.10. LED or laser diode array and SLM. (a) LED or laser diode array for all neurons. (b) SLM for interconnection strengths and bias inputs.

in the storage. To calculate (7.33), the light emitted by the laser diode array is collected vertically and then spread to a spatial light module (SLM) which represents the interconnection strengths and the bias inputs shown in Figure 7.10. After collecting the light emerging from each row of the SLM horizontally, an $L^2 \times 1$ photo detector array is used to detect the output. The elements of the output vector $\underline{U}_i$ are then entered in parallel to a threshold array to calculate (7.13). Another threshold array is used for computing the energy changes as soon as the output of the first threshold array becomes available. By feeding all outputs of the first and second threshold arrays to a decision array, the final result, a row of neurons in a new state, is obtained and stored in the ith row of the storage that was chosen by a control. Consequently, the ith row of the laser diode array is updated. In this system only the matrix-matrix (or matrix-vector) product is implemented optically; other computations are done by electronic

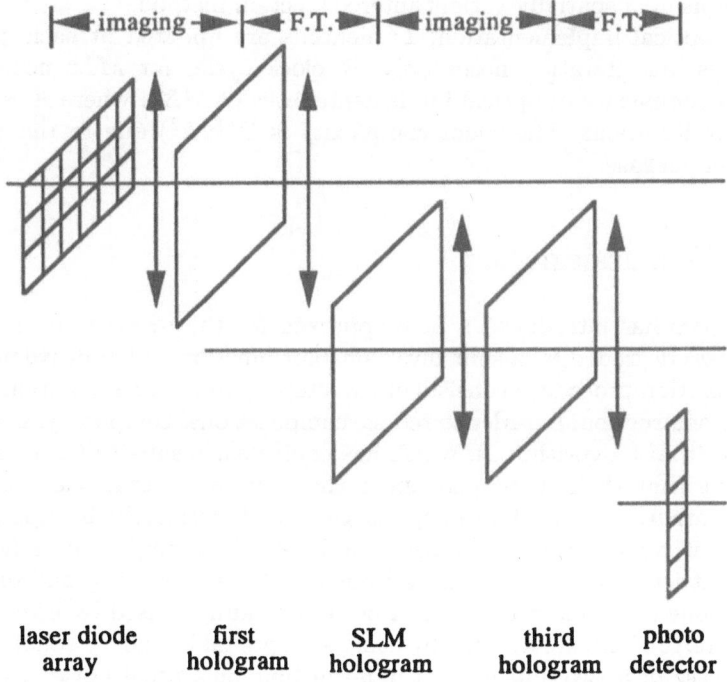

FIGURE 7.11. Schematic of large size matrix-matrix product using holograms.

circuits. To overcome the difficulty of the negative number representation in optics, a coding scheme using two non-negative numbers to represent a bipolar number can be used here [Goo84].

Since the dimensions of the array of neurons and the SLM are very large, we use holograms to implement the matrix-matrix (or matrix-vector) product. It is estimated [SJ86] that with current technology as many as $(10^6 - 10^7)$ gates can be interconnected by using holograms. Figure 7.11 shows a schematic of a large size matrix-matrix product using holograms. The first and third holograms in the figure are computer-generated holograms (CGH) which are used for interconnections to realize the summations in (7.33). The second hologram is used for the SLM to implement multiplication. Supposing that the blur function is spatially invariant, all elements in the same column of the laser diode array are connected to the same $(2 \times K - 1)^2$ elements of the SLM array, where K is the window size of the blur function. Therefore, this is an $M \times L^2$ elements to $(2 \times K - 1)^2 \times L^2$ elements interconnection. The horizontal collection (the second summation at the last step of the matrix-matrix product) is accomplished by the third hologram using a spatially variant interconnection method.

In the optical implementation, L^2 neurons are updated at each clock cycle, thus, one iteration needs only M clock cycles for ML^2 neurons. The time complexity of optical implementaion is $O(MK)$, where K is the number of iterations. The space complexity is $O(ML^2)$ due to the total number of neurons.

7.11 Discussion

This chapter has introduced a new approach for the restoration of gray level images degraded by a shift invariant blur function and additive noise. The restoration procedure consists of two steps: parameter estimation and image reconstruction. In order to reduce computational complexity, a practical algorithm (Algorithm 2), which has equivalent results to the original one (Algorithm 1), is developed under the assumption that the neurons are sequentially visited. The image is generated iteratively by updating the neurons representing the image gray levels via a simple sum scheme. As no matrices are inverted, the serious problems of ringing due to the ill conditioned blur matrix H and of noise overriding caused by inverse or pseudoinverse filters are avoided by using suboptimal boundary conditions. For the case of a two-dimensional uniform blur plus small noise, the approach based on the neural network gives high quality images compared to some of the existing methods. We see from the experimental results that the error defined by (7.16) is small while the error between the original image and the restored image is relatively large. This is because the energy of the neural network is minimized by using (7.16) only. Another reason is that when the blur matrix is singular or ill conditioned, the mapping from $\underline{X}$ to $\underline{Y}$ is not one-to-one, therefore, the error measure (7.16) is not reliable anymore. In our experiments, when the window size of a uniform blur function is 3×3, the ringing effect was eliminated by using blurred noisy boundary values without any smoothness constraint. When the window size is 5×5, the ringing effect was reduced with the help of the smoothness constraint and suboptimal boundary conditions.

8

Conclusions and Future Research

8.1 Conclusions

Several artificial neural networks were presented and applied to computer vision problems such as static and motion stereo, computation of optical flow, and image restoration. To ensure quick convergence of the networks, the deterministic decision rule was used in all the algorithms. Experimental results using natural images confirm that neural networks provide simple but very efficient means to solve computer vision problems, especially at the low-level. Experimental results also provide a strong support to the hypothesis that the first order derivatives of the intensity function and the Gabor features are appropriate measurement primitives for stereo matching, and the principal curvatures are useful for computing optical flow. The utilization of multiple frames for computing depth and optical flow gives much better results and is useful for real time applications. Since no matrices are inverted during restoration, the serious problem of ringing due to the ill conditioned blur matrix is avoided and hence the neural network algorithm gives high quality images compared to some of the existing methods. Although the artificial neural networks have been applied to only a few low-level computer vision problems so far, it is potentially useful for many computer vision problems.

8.2 Future Research

As pointed out in the discussion section of previous chapters, some topics for future study became apparent during the course of this research. How to detect motion discontinuities when the motion is translational and rotational, how to restore blurred and noisy images when the SNR is very low, and how to use the stochastic decision rule to find a global optimal solution to these problems are just a few questions raised in our work. However, the long-term goal of this research is to develop an artificial neural network vision system for the recognition of objects in a three-dimensional scene, which is useful for robot manipulation and vehicle navigation. (A simple example of using Schemas for robot hand control can be found in [AIL87].) The system will make use of parallelism at all levels to achieve real-time vision in complex environments. There exist many vision systems such as ACRONYM [Bro81], 3D MOSAIC [HK84], VISIONS [BRH85], among others, that emphasize different aspects of the three-dimensional object recognition problem. However, none of them are able to match human performance. A typical system has the following structure shown in Figure 8.1. The first part, image restoration, is to pre-process the given digital image to remove system and statistical degradations. The second part is to extract features such as edges, lines, shapes, and optical flow and to segment the image into connected regions that are "homogeneous" in some sense. The third part is to resegment the image based on various geometric criteria, to measure various properties of and relations among the regions, and to establish a relational structure (a labeled scene graph) according to the properties and relations of the image parts (regions). The last part is to match models and to recognize the objects by finding subgraphs of the scene graph satisfying the constraints defined by the object graph. In order to construct such a vision system using neural networks, many interesting and promising research topics should be pursued. Here we discuss a few of them.

For feature extraction, naturally edge detection is the first thing to think of. There are many types of edges, such as step, roof, line, and ramp edges. One possible way to detect edges by using a neural network is to formulate the edge detection problem as a pattern recognition problem. For example, step edges with different orientations may be considered as different patterns. The neural network can then be trained to remember all these patterns. When observations are fed into the network, an edge will be detected if the observation matches one of the patterns.

Although the approach to texture segmentation based on neural networks is not new [HR78], it is still desirable to develop a rotation invariant method. As mentioned in the previous chapter, the human eye is very sensitive to intensity changes. We may use the intensity derivatives to construct some rotation invariant statistics and then segment the image based on such statistics. In addition, segmentation based on geometric structure of surfaces is also a good research topic.

Another important issue is to compute shapes, or Shape, from x. Our static stereo and motion stereo algorithms are for computing shapes from stereo and motion images, respectively. Shape from shading is to find the needle map and then to recover the depth map or directly recover depth without computing the reflectance map. If the needle map is known, a neural network can be used to recover the depth $Z(i,j)$ by minimizing the cost function:

$$\mathbf{E} = \sum_{i=1}^{N_r}\sum_{j=1}^{N_c}\{[\frac{1}{2}(Z(i,j+1) - Z(i,j) + Z(i+1,j+1)$$

$$-Z(i+1,j)) - p(i,j)]^2 + [\frac{1}{2}(Z(i+1,j) - Z(i,j)$$

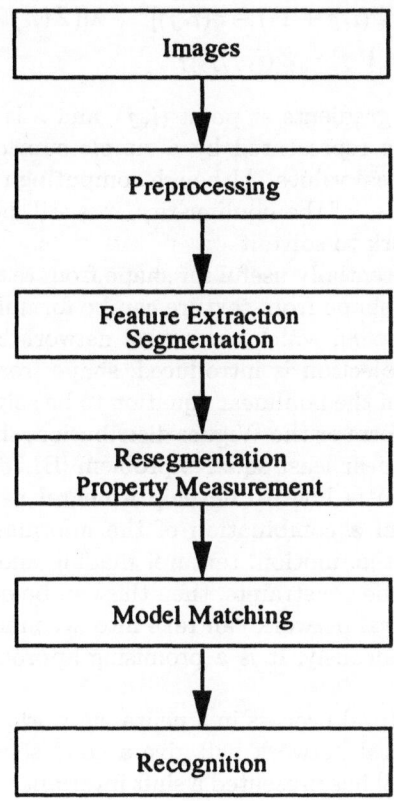

FIGURE 8.1. Vision system structure.

$$+Z(i+1,j+1) - Z(i,j+1)) - q(i,j)]^2 + \lambda[(Z(i,j+1)$$
$$-Z(i,j))^2 + (Z(i+1,j) - Z(i,j))^2]\} \qquad (8.1)$$

where $p(i,j)$ and $q(i,j)$ are the gradients at point (i,j), and λ is a control parameter. The depth $Z(i,j)$ is represented by a simple sum of neuron state variables which may take real values. Although computing the needle map is a difficult problem because of the nonlinearity, it is still possible to use a higher order neural network to solve it.

The neural network is also potentially useful for shape from texture. Under paraperspective projection, shape from texture can be formulated as a linear least squares problem [Alo88], which the neural network is good at handling. When perspective projection is introduced, shape from texture becomes more difficult because of the nonlinear equation to be solved. However, by using the Fourier transform or the Wigner distribution, shape from texture can be simplified as a linear least squares problem [BL76, JC88].

One very challenging problem is how to develop a neural network algorithm to compute shape from a combination of the information from different image cues such as stereo, motion, texture, shading and contour. If one thinks of these cues as some constraints, then this can be done using a neural network. Since the neural network can take into account multiple and mutual constraints simultaneously, it is a promising approach to the integration problem.

Since graph matching is a natural process in a neural network, definitely the approach based on the neural network will give a good solution. Recently, von der Malsburg [vdM88] has presented a shift invariant method for labeled graph matching using a multi-layer neural network. This method uses local features (such as edges, corners and gray level) and the relationships between these features to construct a labeled graph. Distortions such as something being partly hidden are not considered. For three-dimensional matching, the major problem is that the observed image parts may not correspond to the object parts due to changes in perspective such as occlusion and segmentation error (e.g., incorrect merging and splitting). We have to develop an error tolerant matching algorithm to solve this problem. Principally, this algorithm will use object-dependent features to recognize objects based on subgraph matching techniques.

Bibliography

[AA77]   S. Amari and M. A. Arbib. "Competition and cooperation in neural nets". In J. Metzler, editor, *Systems Neuroscience*, pages 119–165. Academic Press, 1977.

[ADG84]   T. D. Albright, R. Desimone, and C. G. Gross. "Columnar organization of directionally selective cells in visual area MT of the Macaque". *J. Neurophysiol.*, vol. 51:16–31, 1984.

[Adi85]   G. Adiv. "Determining three–dimensional motion and structure from optical flow generated by several moving objects". *IEEE Transaction on Pattern Analysis and Machine Intelligence*, vol. PAMI-7:384–401, July 1985.

[AH77]   H. C. Andrews and B. R. Hunt. *Digital Image Restoration*. Prentice-Hall, Englewood Cliffs, New Jersey, 1977.

[AH87]   M. A. Arbib and A. R. Hanson. "Vision, brain, and cooperative computation: An overview". In M. A. Arbib and A. R. Hanson, editors, *Vision, Brain, and Cooperative Computation*, pages 1–83. The MIT Press, 1987.

[AIL87]   M. A. Arbib, T. Iberall, and D. Lyons. "Schemas that integrate vision and touch for hand control". In M. A. Arbib and A. R. Hanson, editors, *Vision, Brain, and Cooperative Computation*, pages 489–510. The MIT Press, 1987.

[Alo88]   J. Aloimonos. "Shape from texture". *Biological Cybernetics*, vol. 58:345–360, 1988.

[Ama71]   S. Amari. "Characteristics of randomly connected threshold-element networks and network systems". *Proceedings IEEE*, vol. 59:35–47, Jan. 1971.

[Ama72]   S. Amari. "Learning patterns and pattern sequences by self-organizing nets of threshold elements". *IEEE Transactions on Computer*, vol. C-21:1197–1206, Nov. 1972.

[Ama77]   S. Amari. "Neural theory of association and concept forma-
           tion". *Biological Cybernetics*, vol. 26:175–185, 1977.

[AMP87]   Y. S. Abu-Mostafa and D. Psaltis. "Optical neural computers".
           Scientific American, pages 88–96, May 1987.

[And77]   J. A. Anderson. "Neural models with cognitive implications".
           In D. LaBerge and S. J. Samuels, editors, *Basic Processes
           in Reading Perception and Comprehension*, pages 27–90. Erl-
           baum, 1977.

[Arb64]   M. A. Arbib. *Brains, Machines, and Mathematics*. McGraw-
           Hill, New York, New York, 1964.

[Bar86]   S. T. Barnard. "A stochastic approach to stereo vision". In
           Proc. Fifth National Conf. on Artificial Intelligence, Philadel-
           phia, PA, August 1986.

[BB85]   R. C. Bolles and H. H. Baker. "Epipolar-plane image analysis:
           A technique for analyzing motion sequences". In *The Proc.
           IEEE Third Workshop on Computer Vision: Representation
           and Control*, pages 168–178, Bellaire, Michigan, 1985.

[BCG90]   A. C. Bovik, M. Clark, and W. S. Geisler. "Multichannel tex-
           ture analysis using localized spatial filters". *IEEE Trans. on
           Patt. Anal. and Mach. Intel.*, vol. PAMI-12:55–73, January
           1990.

[Bec73]   P. Beckmann. *Orthogonal Polynomials for Engineers and
           Physicists*. Golem, Boulder, CO, 1973.

[BH87]   S. D. Blostein and T. Huang. "Quantization errors in stereo
           triangulation". In *Proc. First International Conf. on Com-
           puter Vision*, pages 325–334, London, England, June 1987.

[BL76]   R. Bajcsy and L Lieberman. "Texture gradient as a depth
           cue". *Computer Graphics and Image Processing*, vol. 5:52–67,
           1976.

[BLvdM89] J. Buhmann, J. Lange, and C. von der Malsburg. "Distortion
           invariant object recognition by matching hierarchically labeled
           graphs". In *Proc. Intl. Joint Conf. on Neural Networks*, vol. I,
           pages 155–159, Washington D.C., June 1989.

[BPNA81]   J. F. Baker, S. E. Petersen, W.T. Newsome, and J. M. Allman.
           "Visual response properties of neurons in four extrastriate vi-
           sual areas of the owl monkey (aotus trivirgatus): A quantita-
           tive comparison of medial, dorsomedial, and middle temporal
           areas". *J. Neurophysiol.*, vol. 45:397–416, 1981.

[BRG83]   J. Biemond, J. Rieske, and J. Gerbrand. "A fast Kalman filter for images degraded by both blur and noise". *IEEE Trans. Acoustics,Speech,Signal Processing*, vol. ASSP-31:1248–1256, October. 1983.

[BRH85]   R. Belknap, E. Riseman, and A. Hanson. "The information fusion problem and rule-based hypotheses applied to complex aggregations of image events". In *Proc. DARPA Image Understanding Workshop*, Miami Beach, FL, 1985.

[Bro81]   R. A. Brooks. *Model-Based Computer Vision*. M.I.T. Press, Cambridge, MA, 1981.

[BTBW77] H. G. Barrow, J. M. Tenenbaum, R.C. Bolles, and H. C. Wolf. "Parametric correspondence and chamfer matching: Two new techniques for image matching". In *Proc. Fifth International Joint Conf. on Artificial Intelligence*, Cambridge, MA, 1977.

[CJ85]   R. Chellappa and H. Jinchi. "A nonrecursive filter for edge preserving image restoration". In *Proc. Intl. Conf. on Acoustics, Speech, and Signal Processing*, pages 652–655, Tampa, FL, March 1985.

[CK82]   R. Chellappa and R. L. Kashyap. "Digital image restoration using spatial interaction models". *IEEE Trans. Acoustics, Speech, and Signal Processing*, vol. ASSP-30:461–472, June. 1982.

[CWK82]   S. Lin C, R. E. Weller, and J. H. Kaas. "Cortical connections of striate cortex in the owl monkey". *J. Comp. Neurol.*, vol. 211:165–176, 1982.

[Dau85]   J. G. Daugman. "Uncertainty relation for resolution in space, spatial frequency, and orientation optimized by two-dimensional visual cortical filters". *Journal of the Optical Society of America A*, vol. 2:1160–1169, 1985.

[Dau88]   J. G. Daugman. "Complete discrete 2-D Gabor transforms by neural networks for image analysis and compression". *IEEE Trans. on Acoustic, Speech, and Signal Processing*, vol. ASSP-36:1169–1179, July 1988.

[FB82]   J. A. Feldman and D. H. Ballard. "Connectionist models and their properties". *Cognitive Science*, vol. 6:205–254, 1982.

[FPPP85]   N. H. Farhat, D. Psaltis, A. Prata, and E. Paek. "Optical implementation of the Hopfield model". *Applied Optics*, vol. 24, No.10:1469–1475, 15 May 1985.

[FT79]   C. L. Fennema and W. B. Thompson. "Velocity determina-
           tion in scene containing several moving objects". *Computer
           Graphics and Image Processing*, vol. 9:301–315, 1979.

[Fuk75]   K. Fukushima. "Cognitron: A self-organizing multilayered neu-
           ral network". *Biological Cybernetics*, vol. 20:121–136, 1975.

[Gab46]   D. Gabor. "Theory of communication". *Journal of the Institute
           of Electrical Engineers*, vol. 93:429–459, 1946.

[GG84]   S. Geman and D. Geman. "Stochastic relaxation, Gibbs distri-
           butions, and Bayesian restoration of images". *IEEE Trans. on
           Patt. Anal. and Mach. Intel.*, vol. PAMI-6:141–149, November
           1984.

[GG87]   A. F. Gmitro and G. R. Gindi. "Optical neurocomputer for
           implementation of the Marr–Poggio stereo algorithm". In *Proc.
           IEEE First Annual Intl. Conf. on Neural Networks*, San Diego,
           CA, June 1987.

[GL86]   N. M. Grzywacz and A. L.Yuille. "Motion correspondence
           and analog networks". In *Proc. Conf. on Neural Networks
           for Computing*, pages 200–205, Snowbird, UT, 1986. American
           Institute of Physics.

[GL87]   N. M. Grzywacz and A. L.Yuille. Massively parallel implemen-
           tations of theories for apparent motion. Technical Report AI
           Memo No. 888, CBIP Memo No. 016, MIT Artificial Intelli-
           gence Lab. and Center for Biological Information Processing,
           June 1987.

[GN87]   M. A. Gennert and S. Negahdaripour. Relaxing the brightness
           constancy assumption in computing optical flow. Technical
           Report AI Memo No. 975, MIT Artificial Intelligence Lab.,
           June 1987.

[Goo84]   J. W. Goodman. "The optical data processing family tree".
           Opt. News, vol. 25, 1984.

[Gri81]   W. E. L. Grimson. *From Images to Surfaces*. The MIT press,
           Cambridge, MA, 1981.

[Gri85]   W. E. L. Grimson. "Computational experiments with a fea-
           ture based stereo algorithm". *IEEE Trans. on Patt. Anal. and
           Mach. Intel.*, vol. PAMI-7:17–34, January 1985.

[Gro76]   S. Grossberg. "Adaptive pattern calssification and universal
           recording: Part I. parallel development and coding of neural
           feature detectors". *Biological Cybernetics*, vol. 23:121–134,
           1976.

[HA87]   W. Hoff and N. Ahuja. "Extracting surfaces from stereo images: An integrated approach". In *Proc. First International Conf. on Computer Vision*, pages 284–294, London, England, June 1987.

[Had02]   J. Hadamard. "Sur les problèmes aux dérivées partielles et leur signification physique". *Princeton University Bulletin*, vol. 13, 1902.

[Har84]   R. M. Haralick. "Digital step edges from zero crossings of second directional derivatives". *IEEE Trans. Patt. Anal. Mach. Intell.*, vol. PAMI-6:58–68, January 1984.

[Heb49]   D. O. Hebb. *The Organization of Behavior*. Wiley, New York, 1949.

[Hil83]   E. C. Hildreth. *The Measurement of Visual Motion*. The MIT Press, Cambridge, Massachusetts, 1983.

[HK84]   M. Herman and T Kanade. "The 3D mosaic scene understanding system: Incremental reconstruction of 3D scenes from complex images". In *Proc. DARPA Image Understanding Workshop*, Norwood, NJ, 1984.

[HKLM88]   J. Hutchinson, C Koch, J Luo, and C. Mead. "Computing motion using analog and binary resistive networks". *IEEE Computer Magazine*, pages 52–63, March 1988.

[Hop82]   J. J. Hopfield. "Neural networks and physical systems with emergent collective computational abilities". *Proc. Natl. Acad. Sci. USA*, vol. 79:2554–2558, April 1982.

[Hop84]   J. J. Hopfield. "Neurons with graded response have collective computational properties like those of two-state neurons". *Proc. Natl. Acad. Sci. USA*, vol. 81:3088–3092, May 1984.

[Hor86]   B. K. P. Horn. *Robot Vision*. The MIT Press, Cambridge, Massachusetts, 1986.

[HR78]   A. R. Hanson and E. M. Riseman. "Segmentation of natural scenes". In A. R. Hanson and E. M. Riseman, editors, *Computer Vision Systems*. Academic Press, 1978.

[HS81]   B. K. P. Horn and B. G. Schunck. "Determining optical flow". *Artificial Intelligence*, vol. 17:185–203, August 1981.

[HT85]   J. J. Hopfield and D. W. Tank. "Neural computation of decisions in optimization problems". *Biological Cybernetics*, vol. 52:141–152, 1985.

[HW65]   D. H. Hubel and T. N. Wiesel. "Receptive fields and functional architecture in two nonstriate visual areas (18 and 19) of the cat". *J. Neurophysiol.*, vol. 128:229–289, 1965.

[HW77]   D. H. Hubel and T. N. Wiesel. "Functional architecture of Macaque monkey visual cortex". *Proc. R. Soc. Lond. Ser. B*, vol. 198:1–59, 1977.

[IMO84]   H. Itoh, A Miyauchi, and S. Ozawa. "Distance measuring method using only simple vision constrained for moving robots". In *Seventh Intl. Conf. Pattern Recognition*, pages 192–195, Montreal, July 1984.

[JBO87]   R. Jain, S. L. Bartlett, and N. O'Brien. "Motion stereo using ego-motion complex logarithmic mapping". *IEEE Trans. on Patt. Anal. and Mach. Intel.*, vol. PAMI-9:356–369, May 1987.

[JC86]   H. Jinchi and R. Chellappa. "Restoration of blurred and noisy image using Gaussian Markov random field models". In *Proc. Conf. on Information Science and System*, pages 34–39, Princeton Univ., NJ, 1986.

[JC88]   Y. C. Jau and R. T. Chin. "Shape from texture using the Wigner distribution". In *The Proc. IEEE Conf. on Computer Vision and Pattern Recognition*, pages 515–523, Ann Arbor, MI, 1988.

[Jul60]   B. Julesz. "Binocular depth perception of computer–generated patterns". *Bell System Technical J.*, vol. 39:1125–1162, Sept. 1960.

[Jul71]   B. Julesz. *Foundations of Cyclopean Perception.* The University of Chicago Press, Chicago, IL, 1971.

[KGV83]   S. Kirkpatrick, C. D. Gelatt, and M. P. Vecchi. "Optimization by stimulated annealing". *Science*, vol. 220:671–680, 1983.

[Koc87]   C. Koch. "Analog neuronal networks for real-time vision systems". In *Proc. Workshop on Neural Network Devices and Applications*, Los Angeles, CA, February 1987.

[Koh77]   T. Kohonen. *Associative Memory: A System Theoretical Approach.* Springer-Verlag, New York, NY, 1977.

[Kos88]   B. Kosko. "Bidirectional associative memories". *IEEE Trans. Systems, Man, and Cybernetics*, vol. SMC-18:42–60, 1988.

[LaS86]   J. P. LaSalle. *The Stability and Control of Discrete Processes.* Springer-Verlag, New York, NY, 1986.

[LHW82]   T. J. Laffey, R. M. Haralick, and L. T. Watson. "Topographic classification of digital image intensity surfaces". In *The Proc. IEEE Workshop on Computer Vision: Theory and Control,* Rindge, New Hampshire, August 1982.

[LM75]   J. O. Limb and J. A. Murphy. "Estimating the velocity of moving objects from television signals". *Computer Graphics and Image Processing,* vol. 4:311–329, 1975.

[Lor66]   G. G. Lorentz. *Approximation of functions.* Holt, Rinehart and Winston, New York, 1966.

[Mar82]   D. Marr. *Vision.* W. H. Freeman and Company, New York, 1982.

[MF81]   J. E. W. Mayhew and J. P. Frisby. "Psychophysical and computational studies towards a theory of human stereopsis". *Artificial Intelligence,* vol. 17:349–385, August. 1981.

[MH80]   D. Marr and E. C. Hildreth. "Theory of edge detection". *Proc. Royal Society of London,* vol. B-207:187–217, February 1980.

[MN85]   G. Medioni and R. Nevatia. "Segment–based stereo matching". *Computer Vision, Graphics, and Image Processing,* vol. 31:2–18, 1985.

[Mon80]   V. M. Montero. "Patterns of connections from striate cortex to cortical visual area in superior temporal sulcus of Macaque and middle temporal gyrus of owl monkey". *J. Comp. Neurol.,* vol. 189:45–59, 1980.

[Mor84]   V. A. Morozov. *Methods for Solving Incorrectly Posed Problems.* Springer-Verlag, New York, 1984.

[MP43]   W. S. McCulloch and W. Pitts. "A logical calculus of the ideas immanent in nervous activity". *Bulletin of Mathematical Biophysics,* vol. 5:115–133, 1943.

[MP69]   M. Minsky and S. Papert. *Perceptrons.* MIT Press, Cambridge, 1969.

[MP76]   D. Marr and T. Poggio. "Cooperative computation of stereo disparity". *Science,* vol. 194:283–287, October 1976.

[MPH79]   D. Marr, T. Poggio, and E. C. Hildreth. "The smallest channel in early human vision". *Journal of the Optical Society of America,* vol. 90:868–870, 1979.

[MRR+53]   N. Metropolis, A. W. Rosenbluth, M. N. Rosenbluth, A. H. Teller, and E. Teller. "Equations of state calculations by fast computing machines". *J. Chem. Phys.*, vol. 21:1087–1091, 1953.

[MSK88]   L. Matthies, R. Szeliski, and T Kanade. "Kalman filter-based algorithms for estimating depth from image sequences". In *Proc. DARPA Image Understanding Workshop*, pages 199–213, Cambridge, MA, April 1988.

[Nag83]   H. H. Nagel. "Displacement vectors derived from second order intensity variations in image sequence". *Computer Vision, Graphics, and Image Processing*, vol. 21:85–117, 1983.

[NB80]   R. Nevatia and K. R. Babu. "Linear feature extraction and description". *Computer Graphics and Image Processing*, vol. 13:257–269, 1980.

[Nev76]   R. Nevatia. "Depth measurement by motion stereo". *Computer Graphics and Image Processing*, vol. 6:619–630, 1976.

[NR79]   A. N. Netravali and J. D. Robbins. "Motion-compensated television coding: Part I". *The Bell System Technique Journal*, vol. 58:631–670, 1979.

[NR80]   A. N. Netravali and J. D. Robbins. "Motion-compensated coding: Some new results". *The Bell System Technique Journal*, vol. 59:1735–1745, 1980.

[O'N66]   B. O'Neill. *Elementary Differential Geometry*. Academic Press, New York, NY, 1966.

[PA83]   J. M. Prager and M. A. Arbib. "Computing the optical flow: The match algorithm and prediction". *Computer Vision, Graphics, and Image Processing*, vol. 24:271–304, 1983.

[PD75]   P. Dev. "Perception of depth surfaces in random-dot stereogram: a neural model". *Int. J. Man-Machine Studies*, 7:511–528, 1975.

[PFG78]   W. K. Pratt, O. D. Faugeras, and A. Gagalowicz. "Visual discrimination of stochastic texture fields". *IEEE Trans. Syst., Man, Cybern.*, vol. SMC-8:796–814, Nov. 1978.

[Pog84]   T. Poggio. "Vision by man and machine". *Scientific American*, vol. 250:106–116, April. 1984.

[Pot75]   J. L. Potter. "Velocity as a cue to segmentation". *IEEE Trans. Systems, Man, and Cybernetics*, vol. SMC-5:390–394, 1975.

[Pra78]   W. K. Pratt. *Digital Image Processing*. John Wiley and Sons, New York, 1978.

[Pre70]   J. M. Prewitt. "Object enhancement and extraction". In B. S. Lipkin and A. Rosenfeld, editors, *Picture Processing and Psychopictorics*. Academic Press, New York, 1970.

[PTK85]   T. Poggio, V. Torre, and C. Koch. "Computational vision and regularization theory". *Nature*, vol. 317:314–319, Sept. 1985.

[PZ88]   M. Porat and Y. Y. Zeevi. "The generalized gabor scheme of image representation in biological and machine vision". *IEEE Trans. on Patt. Anal. and Mach. Intel.*, vol. PAMI-10:452–468, July 1988.

[RM82]   D. E. Rumelhart and J. L. McClelland. "An interactive activation model of context effects in letter perception: Part 2. The contextual enhancement effect and some tests and extensions of the model". *Psychological Review*, vol. 89:75–112, 1982.

[Ros59]   F. Rosenblett. "Two theorems of statistical separability in the perceptron". In *Mechanisation of Thought Processes: Proceedings of a Symposium Held at the National Physical Laboratory*, pages 421–456. HM Stationery Office, London, November, 1959.

[Ros62]   F. Rosenblett. *Principles of Neurodynamics*. Spartan, New York, 1962.

[SCL87]   G. Z. Sun, H. H. Chen, and Y. C. Lee. "Learning stereopsis with neural networks". In *Proc. IEEE First Annual Intl. Conf. on Neural Networks*, San Diego, CA, June 1987.

[SD87]   C. V. Stewar and C. R. Dyer. "A connectionist model for stereo vision". In *Proc. IEEE First Annual Intl. Conf. on Neural Networks*, San Diego, CA, June 1987.

[Sel55]   O. G. Selfridge. "Pattern recognition in modern computers". In *Proc. of the Western Joint Computer Conference*, 1955.

[SJ86]   A. A. Sawchuk and B. K. Jenkins. "Dynamic optical interconnections for parallel processors". In *Proc. of Optical Computing Symposium on Optoelectronics and Laser Applications in Science and Engineering*, SIPE 625, Los Angeles, CA, January 1986.

[Son72]   M. M. Sondhi. "The removal of spatially invariant degradations". *Proc. of IEEE*, vol. 60:842–853, July 1972.

[SR86]   J. Sejnowski and C. R. Rosenberg. Nettalk: A parallel network that learns to read aloud. Technical Report JHU-EECS-86-01, Johns Hopkins Univ., 1986.

[TG86]   M. Takeda and J. W. Goodman. "Neural networks for computation: Number representations and programming complexity". *Applied Optics*, vol. 25, No. 18:3033–3046, Sept. 1986.

[TH87]   D. W. Tank and J. J. Hopfield. "Concentrating information in time: Analog neural networks with possible applications to speech recognition". In *Proc. IEEE First Annual Intl. Conf. on Neural Networks*, vol. IV, pages 455–468, San Diego, CA, June 1987.

[Tho80]   W. B. Thompson. "Combining motion and contrast for segmentation". *IEEE Trans. on Patt. Anal. and Mach. Intel.*, vol. PAMI-2:543–549, 1980.

[Tho81]   W. B. Thompson. "Lower-level estimation and interpretation of visual motion". *Computer*, pages 20–28, August 1981.

[TMB85]   W. B. Thompson, K. M. Mutch, and V. A. Berzins. "Dynamic occlusion analysis in optical flow fields". *IEEE Trans. on Patt. Anal. and Mach. Intel.*, vol. PAMI-7:374–383, July 1985.

[TTA+81]   J. Tiggs, M. Tiggs, S Anschel, N.A. Cross, W. D. Ledbetter, and R. L. McBride. "Areal and laminar distribution of neurons interconnecting the central visual cortical areas 17, 18 19 and MT in squirrel monkey (saimiri)". *J. Comp. Neurol.*, vol. 202:539–560, 1981.

[Ull79]   S. Ullman. *The Interpretation of Visual Motion*. M.I.T. Press, Cambridge, MA, 1979.

[vdM73]   C. von der Malsburg. "Self-organization of orientation sensitive cells in the striate cortex". *Kybernetik*, vol. 14:85–100, 1973.

[vdM88]   C. von der Malsburg. "Pattern recognition by labeled graph matching". *Neural Networks*, vol. 1, Number 2:141–148, 1988.

[WB79]   H. R. Wilson and J. R. Bergen. "A four mechanism model for threshold spatial vision". *Vision Research*, 19:19–32, 1979.

[WBT85]   J. W. Woods, J. Biemond, and A. M. Tekalp. "Boundary value problem in image restoration". In *Proc. Intl. Conf. on Acoustics, Speech, and Signal Processing*, pages 692–695, Tampa, FL, March 1985.

[WG77]   H. R. Wilson and S. C. Giese. "Threshold visibility of frequency grating patterns". *Vision Research*, 17:1177–1190, 1977.

[WH60]   B. Widrow and M. E. Hoff. "Adaptive switching circuits". In *Institute of Radio Engineers, Western Electronic Show and Convention, Convention Record*, vol. Part 4, pages 96–104, August 1960.

[Whe38]   Sir C. Wheatstone. "Contribution to the physiology of vision". *Phil. Trans. R. Soc.*, 1938.

[WI81]   J. W. Woods and V. K. Ingle. "Kalman filtering in two dimensions: Further results". *IEEE Trans. Acoustics,Speech,Signal Processing*, vol. ASSP-29:188–197, April. 1981.

[Wid59]   B. Widrow. "Adaptive sampled-data systems–a statistical theory of adaptation ". In *Institute of Radio Engineers, Western Electronic Show and Convention, Convention Record*, vol. Part 4, pages 74–85, 1959.

[Wil80]   T. Williams. "Depth from camera motion in a real world scene". *IEEE Trans. on Patt. Anal. and Mach. Intel.*, vol. PAMI-2:511–516, November 1980.

[Win77]   P. Wintz. *Digital Image Processing*. Addison-Wesley, Reading, Massachusetts, 1977.

[WS85]   B. Widrow and S. Stearns. *Adaptive Signal Processing*. Prentice-Hall, Englewood Cliffs, 1985.

[XTA87]   G. Xu, S. Tsuji, and M. Asada. "A motion stereo method based on coarse-to-fine control strategy". *IEEE Trans. on Patt. Anal. and Mach. Intel.*, vol. PAMI-9:332–336, March 1987.

[ZC88a]   Y. T. Zhou and R. Chellappa. "A neural network approach to stereo matching". In *Proc. Applications of Digital Image Processing* XI, SPIE. 974, pages 243–250, San Diego, CA, August 1988.

[ZC88b]   Y. T. Zhou and R. Chellappa. "Stereo matching using a neural network". In *Proc. Intl. Conf. on Acoustics, Speech, and Signal Processing*, pages 940–943, New York, NY, April 1988.

[ZC89]   Y. T. Zhou and R. Chellappa. "Neural network algorithms for motion stereo". In *Proc. Intl. Joint Conf. on Neural Networks*, vol. 2, pages 251–258, Washington D.C., June 1989.

[ZC90]   Y. T. Zhou and R. Chellappa. "A network for motion percep-
            tion". In *Proc. Intl. Joint Conf. on Neural Networks*, vol. 2,
            pages 875–884, San Diego, CA., June 1990.

[ZC91]   Y. T. Zhou and R. Crawshaw. "Contrast, size and orientation
            invariant target detection in infrared imagery". In *Proc. Appli-
            cations of Automatic Object Recognition*, SPIE. 1471, Orlando,
            Florida, April 1991.

[ZCVJ88]   Y. T. Zhou, R. Chellappa, A. Vaid, and B. K. Jenkins. "Im-
            age restoration using a neural network". *IEEE Trans. Acous-
            tics,Speech,Signal Processing*, vol. 36:1141–1151, July 1988.

[ZG90]   Y. Y. Zeevi and R. Ginosar. "Neural computers for foveat-
            ing vision systems". In R. Eckmiller, editor, *Advanced Neural
            Computers*, pages 323–330. Elsevier Science Publishers B.V.,
            North-Holland, 1990.

Index